Thermodynamik

Energie • Umwelt • Technik

Band 33

λογος

Thermodynamik

Energie • Umwelt • Technik

Herausgegeben von Professor Dr.-Ing. Dieter Brüggemann
Ordinarius am Lehrstuhl für Technische Thermodynamik und
Transportprozesse (LTTT) der Universität Bayreuth

Multiscale thermo-fluid modelling of macro-encapsulated latent heat thermal energy storage systems

Von der Fakultät für Ingenieurwissenschaft
der Universität Bayreuth
zur Erlangung der Würde eines
Doktor-Ingenieurs (Dr.-Ing.)
genehmigte Dissertation

von
Raghavendra Rohith Kasibhatla, M.Sc.
aus
Hyderabad, Indien

Erstgutachter: Prof. Dr.-Ing. Dieter Brüggemann
Zweitgutachterin: Prof. Dr. rer. nat. habil. Cornelia Breitkopf
Tag der mündlichen Prüfung: 23.11.2018

Lehrstuhl für Technische Thermodynamik und Transportprozesse (LTTT)
Universität Bayreuth
2018

Thermodynamik: Energie, Umwelt, Technik
Herausgegeben von Prof. Dr.-Ing. D. Brüggemann

Raghavendra Rohith Kasibhatla:
Multiscale thermo-fluid modelling of macro-encapsulated latent heat thermal energy storage systems;
Bd. 33 der Reihe: D. Brüggemann (Hrsg.), „Thermodynamik: Energie, Umwelt, Technik";
Logos-Verlag, Berlin (2019)
zugleich: Diss. Univ. Bayreuth, 2018

Bibliografische Information der Deutschen Nationalbibliothek

Die Deutsche Nationalbibliothek verzeichnet diese Publikation in der Deutschen Nationalbibliografie; detaillierte bibliografische Daten sind im Internet über http://dnb.d-nb.de abrufbar.

ISSN 1611-8421
ISBN 978-3-8325-4893-3

Logos Verlag Berlin GmbH
Comeniushof, Gubener Str. 47,
10243 Berlin
Tel.: +49 030 42 85 10 90
Fax: +49 030 42 85 10 92
INTERNET: http://www.logos-verlag.de

Vorwort des Herausgebers

Thermische Energiespeicher (TES), für die auch der Begriff „Wärmespeicher“ gebräuchlich ist, sind immer dann besonders interessant, wenn Angebot und Nachfrage an Energie variieren und zeitlich nicht deckungsgleich sind. Solche Situationen treten besonders häufig durch die zunehmende Umstellung auf erneuerbare Energieformen und deren vorrangige Einspeisung auf, da Wind und Sonne naturgemäß nur sehr unstetig Energie liefern. Zudem kann auch Abwärme durch Speicherkonzepte besser genutzt werden.

Am Lehrstuhl für Technische Thermodynamik und Transportprozesse (LTTT) im Zentrum für Energietechnik (ZET) der Universität Bayreuth wurden und werden solche Konzepte in mehreren Projekten für verschiedene stationäre und mobile Anwendungen entwickelt und bewertet. Hierbei wird sowohl die sensible Speicherung untersucht, bei der sich mit der aufgenommenen oder abgegebenen Energie die Temperatur fühlbar verändert, wie auch die latente Speicherung, bei der diese sich in einem Phasenübergang verbirgt. Ein weiteres Unterscheidungsmerkmal besteht darin, ob das Speichermaterial weitgehend frei oder speziell verkapselt ist. Letzteres bedeutet mehr Aufwand, eröffnet aber auch Vorteile bei der Wärmeübertragung. Gerade bei vielen phase change materials (PCM) ist nämlich nicht die speicherbare Energie, sondern die Leistung bei der Be- und Entladung aufgrund einer mäßigen Wärmeleitfähigkeit für Anwendungen begrenzend.

Der Autor hat sich intensiv mit Phänomenen beschäftigt, wie sie bei makroverkapselten PCM-Speichern auftreten. Er hat hierzu Modelle entwickelt, um auf den unterschiedlichen räumlichen Skalen einerseits einzelne Kapseln und andererseits gesamte Speicher mit vielen solcher Kapseln zu beschreiben. Die Modelle, deren Umsetzung und ein Vergleich der Simulationsergebnisse mit experimentell gewonnenen Messdaten stellt er hier vor.

Bayreuth, im Februar 2019 — Professor Dr.-Ing. Dieter Brüggemann

Abstract

Energy storage systems shall compensate the inevitable fluctuations of renewable energy sources as well as increase energy efficiency, for example by a better utilization of waste heat. In this context, latent heat thermal energy storage (LHTES) units act as both seasonal and short period storages. However, the poor thermal conductivity of phase change materials (PCM) in LHTES units may result in low charging and discharging powers. This can be improved by macro-encapsulation of PCM.

In framework of the present work, numerical modelling and simulation of macro-encapsulated LHTES units at multiple length scales are discussed and performed. A single computational fluid dynamics (CFD) platform is employed to represent both detailed and simplified models for macro-encapsulated LHTES units. Here, detailed multiphase fluid dynamics and heat transfer inside a PCM capsule are presented with a multiphase numerical model. A simplified convective model accounts for the air phase in PCM capsule as convective boundary condition. Similarly, the simplified conductive model compromises the detailed representation of close contact melting and natural convection with an effective thermal conductivity correlation. Detailed heat transfer and fluid flow in macro-encapsulated LHTES units are modelled with a conjugate heat transfer (CHT) model in this work. The local thermal non-equilibrium (LTNE) model employs volume average equations to represent the heat transfer and fluid flow in a macro-encapsulated LHTES unit.

A comparison of the results from the numerical models with experiments has shown a very good agreement. The developed numerical models may serve as a useful tool for optimizing LHTES enabling the study of different storage configurations within an acceptable time and thus extending the application spectrum of LHTES units for different utilizations.

Kurzfassung

Energiespeicher sollen zum einen die bei erneuerbaren Energien wie Sonne und Wind unvermeidbaren zeitlichen Schwankungen des Angebots ausgleichen und zum anderen die Energieeffizienz, zum Beispiel durch die bessere Nutzung von Abwärme, erhöhen. In diesem Kontext können latente thermische Speicher (LTS) als saisonale oder Kurzzeitspeicher eingesetzt werden. Jedoch führt die geringe Wärmeleitfähigkeit der in LTS eingesetzten Phasenwechselmaterialien (PCM) zu geringen Be- und Entladeleistungen des Speichers. Eine Verbesserung ist durch die Makroverkapselung der PCM möglich.

Im Rahmen dieser Arbeit werden eine numerische Modellierung und die Simulation von makroverkapselten LTS verschiedener Größenskalen diskutiert und durchgeführt. Ein numerisches Mehrphasenmodell zeigt hierbei die detaillierten strömungsmechanischen Vorgänge und die Wärmeübertragung innerhalb einer Makrokapsel. Ein daraus abgeleitetes vereinfachtes Konvektionsmodell berücksichtigt die Luftphase in der Makrokapsel als konvektive Randbedingung. Ein noch weiter vereinfachtes Modell berücksichtigt das Kontaktschmelzen und die natürliche Konvektion mithilfe einer Korrelation der effektiven Wärmeleitfähigkeit. Die Modellierung der detaillierten Wärmeübertragung und der Durchströmung des makroverkapselten LTS erfolgt mit einem „conjugate heat transfer (CHT)"-Modell. Das hierfür verwendete „local thermal non-equilibrium (LTNE)"-Modell nutzt über das Volumen gemittelte Gleichungen, um die Wärmeübertragung und die Durchströmung des makroverkapselten LTS zu beschreiben.

Die Ergebnisse der numerischen Modelle zeigen sehr gute Übereinstimmung mit den Experimenten. Die entwickelte Modellierung hat das Potential als Hilfsmittel zur Optimierung von LTS zu dienen, indem unterschiedliche Speicherkonfigurationen in überschaubarer Zeit untersucht werden können und sich somit der Einsatzbereich von LTS auf unterschiedlichste Anwendungsbereiche erweitern lässt.

Contents

Nomenclature

Symbols

$\mathbf{A}$	Discretized coefficient matrix	-
A	Exponential coefficient for viscosity correlation	-
A_{FW}	External surface area of capsule	m^2
A_{WM}	Internal surface area of capsule	m^2
a	Discretized energy equation coefficient	-
a_{sf}	Specific surface area	1/m
b	Discretized energy equation coefficient	-
Co	Courant number	-
C	Carman-Kozeny constant	kg/m^3s
C_{ht}	Heat transfer coefficient	W/m^2K
c^A	Mass specific apparent heat capacity	m^2/s^2K
c_L	Specific heat capacity in liquid phase	m^2/s^2K
c_S	Specific heat capacity in solid phase	m^2/s^2K
c^{sen}	Specific heat capacity in sensible part	m^2/s^2K
D	Darcy source term	-
D_c	Morphological constant in Darcy term	kg/m^3s
$\mathbf{D}_{ij}$	Deformation tensor	m
D_s	Zero non-divisional constant in Darcy term	-
d	Discretized energy equation coefficient	-
d_p	Particle diameter	m
erf	Error function	-
$\mathbf{F}$	Force	kgm/s^2
F	Forchheimer constant	-
Fo	Fourier number	-
G	Grid	-
Gr	Grashof number	-
Gz	Graetz number	-
$\mathbf{g}$	Gravitational acceleration	m/s^2
H	Heaviside step function	-
h	Mass specific enthalpy	m^2/s^2
K	Permeability	m^2
L	Latent heat	m^2/s^2
l	Length	m
$\mathbf{M}$	Momentum	kgm/s
m	Mass	kg
$\dot{m}$	Mass flow rate	kg/s

Nu	Nusselt number	-
$\mathbf{n}$	Normal vector	-
Pe	Peclet number	-
Pr	Prandtl number	-
p	Pressure	kg/ms^2
p_d	Dynamic pressure	kg/ms^2
$\mathbf{Q}$	Discretized source matrix	-
Q	Heat flow	W
$\dot{Q}$	Heat flux	W/m^2
R	Thermal resistance	K/W
Ra	Rayleigh number	-
Re	Reynolds number	-
r_i	Inner radius	m
r_o	Outer radius	m
S	Source term	-
S_f	Cell surface area	m^2
Ste	Stefan number	-
T	Temperature	K
t	Time	s
tanh	Hyperbolic tan function	-
$\tanh^{-1}$	Inverse tan hyperbolic function	-
Δt	Time step	s
$\mathbf{u}$	Velocity vector	m/s
$\langle \mathbf{u} \rangle$	Local volume averaged velocity	m/s
$\mathbf{u}_r$	Relative velocity vector	m/s
V	Volume of the cell	m^3
$\dot{V}$	Volumetric flow rate	m^3/s
Δx	Finite volume length	m

Greek alphabets

α	Volume fraction of any phase	-
β	Thermal expansion coefficient	1/K
Γ	Diffusivity	-
γ	Porosity or Volume fraction of phase	-
ε	Porosity	-
ε_{conv}	Criterion for convergence	-
θ	Contact angle	$\circ$
κ	Thermal conductivity	W/(mK)
κ_{sf}	Curvature	1/m
λ	Enthalpy jump in a melting range	m^2/s^2
μ	Dynamic viscosity of the fluid	m^2/s
ν	Kinematic viscosity of the fluid	kg/ms
ξ	Drag coefficient	-

ρ	Density	kg/m^3
σ	Surface tension coefficient	N/m
σ_m	Square root of variance in Gaussian function	K
τ	Shear stress	kg/ms^2
ϕ	Scalar quantity	-
φ	Physical quantity	-
χ	Weight fraction of capsule wall in a PCM capsule	-
ψ	Compressibility	s^2/m^2
ω	Under relaxation factor	-

Abbreviations

BBO	Basset Bousinessq Oseen
CFD	Computational fluid dynamics
CFL	Courant Friedrich Lewy
CLSVOF	Coupled level set and volume of fluid
CV	Control volume
HTF	Heat transfer fluid
FDM	Finite difference method
FEM	Finite element method
FVM	Finite volume method
LHTES	Latent heat thermal energy storage
LS	Level set
LTNE	Local thermal non-equilibrium
OpenFOAM	Open field operation and manipulation
PCM	Phase change material
PISO	Pressure implicit with splitting of operators
VOF	Volume of fluid
VVM	Variable viscosity method

Subscripts

Air	Air phase
B	Packed bed
Cap	Capsule
CV	Control volume
c	Cold
eff	Effective
F	Fluid domain
h	Hot
L	Liquid
M	PCM domain
m	Melting point
mom	Momentum

P	Node of the computing cell
PCM	PCM phase
S	Solid
sf	Surface force
W	Wall domain
w	Wall
∞	Ambient

Superscripts

m	Old iteration level
$m+1$	Actual iteration level
nb	Neighbouring cells
o	Old time step
oo	Post-previous old time step
$*$	Uninfluenced of pressure gradient

1 Introduction

The Paris agreement was approved by approximately 170 countries, which establish goals to reduce global temperature by 1.5 to 2 °C [1]. To meet these demands, reducing exhaust gas emissions from conventional energy forms is also equally important alongside moving to renewable energy forms. According to REN21, renewable energy production contributed only up to 20 % of the global energy supply in 2016 [2]. Due to the growing infrastructure and increasing energy demand, conventional energy forms cannot be terminated very soon without sufficient contribution from alternative energy forms. Therefore, reusing energy losses from conventional energy processes is a wise solution, which supports emission reduction alongside increasing system efficiency.

In many industrial processes heat is produced as a bi-product, which is generally gone "wasted" when released into environment. This heat can actually be stored as thermal energy in different forms. Thermal energy storage units in this aspect play an efficient role to store this energy and deliver it back when needed. Thermal energy can stored in the form of sensible heat, latent heat and thermo-chemical heat [3]. Thermo-chemical energy storage (TCES) units involve chemical reactions to store thermal energy, which is influenced on the chemical stability and operating conditions. Sensible heat thermal energy storage (SHTES) units are relevant in processes where waste heat can be instantly required, because they exhibit larger heat losses compared to latent heat thermal energy storage (LHTES) units. In addition, SHTES units also have a lower storage density compared to latent heat thermal energy storage (LHTES) units. Whereas, LHTES units which employ phase change materials (PCM) show high storage density compared to SHTES units.

LHTES units with PCM showcase a high storage capacity but also low charging and discharging powers. The low charging and discharging of LHTES units is triggered by poor thermal conductivity of PCM. In this case, macro-encapsulation of PCM is an effective way to treat the low charging and discharging powers through an enhanced heat transfer area [4]. Meanwhile, macro-encapsulation of PCM offers a wide range of encapsulation forms alongside different layout possibilities. Heat transfer fluid (HTF) used to charge and discharge the storage unit also plays an important role on the heat transfer around the PCM capsule. Therefore, to develop an efficient LHTES unit, different parameters need to be studied. Practical construction of LHTES units with such different configurations is expensive and time consuming. Here, numerical modelling and simulation of macro-encapsulated LHTES unit at different scales provide a detailed overview of the parametric influence on efficiency. Therefore, a detailed study of macro-encapsulated LHTES unit at multiple scales is presented in the present work. Different length scales are defined, where PCM capsule size is considered as the smallest scale and LHTES unit as large scale. The row of multiple PCM capsules in a LHTES unit is defined as an intermediate scale.

To store thermal energy below 100 °C, water is generally affordable. But, to store thermal energy at higher temperatures, PCM are more appropriate. Alongside storing thermal energy at high temperatures, PCM also offer a high storage capacity. This high storage capacity when combined with macro-encapsulation offers a wide range of applications ranging from stationary to mobile storage units. Transport of waste heat with mobile storage units has a great potential to serve energy demands [5]. Thereby, in the present work a laboratory scale macro-encapsulated LHTES unit for mobile applications is developed. For the developed laboratory scale macro-encapsulated LHTES unit, salt-hydrate $MgCl_2 \cdot 6H_2O$ is employed as PCM and aluminium is used as an encapsulation material. Therefore, a great demand to study the behaviour of LHTES unit exists at a large scale. But, due an opaque liquid phase of $MgCl_2 \cdot 6H_2O$, it is difficult to employ optical measurement technique to study phase change in a PCM capsule. Therefore for small and intermediate scales, paraffin RT35HC from Rubitherm is employed to validate numerical models for solid-liquid phase change.

During the charging and discharging of a LHTES unit, melting and solidification of PCM in the PCM capsule is observed respectively. Due to the presence of liquid natural convection and close contact melting, melting of PCM is relatively more complex than solidification process. These effects in a melting process can accelerate and deteriorate the melting rate of PCM. Whereas, PCM solidification in a macro capsule can be represented by a simple Fourier conduction equation. Therefore in the present work, numerical model is employed to investigate and validate a melting process inside a PCM capsule. Whereas, using the complete numerical model for a large scale LHTES unit both charging and discharging processes are studied. Description of numerical models along with the structure of this work is described in chapter 4.

2 Motivation for the present work

In order to meet the increasing energy demands and reducing emissions, installation of latent heat thermal storage units in different industrial applications need to be intensified. For better thermal charging and discharging with large heat transfer surface, macro-encapsulation of PCM is relevant. Moreover, to achieve better storage configurations studying heat transfer and fluid flow inside macro-encapsulated LHTES unit at different scales is also very important. But, due to large choice of PCM and capsule forms it is difficult to investigate various storage parameters of a macro-encapsulated LHTES unit through prototyping. In this case, analytical models are not sufficient to represent the detailed physical occurrence of fluid flow and heat transfer in a LHTES unit accurately. Therefore, numerical modelling and simulation plays a vital role here to understand and optimize the behaviour of thermal storage unit at different length scales.

The goal of this work is to bridge the gaps between detailed modelling of PCM capsule and system simulation of storage units, which supports an effortless parametric study of the thermal storage units. Therefore, in the present work a multi-scale numerical modelling is introduced to represent latent heat thermal storage units at different scales. Its focus answers the following questions:

1. How evolved is the field of numerical simulation to substantially study the behaviour of LHTES units at different length scales?
2. How can physical process in a small scale PCM capsule be represented in a large scale macro-encapsulated LHTES unit?
3. What is the role of natural convection and close contact melting in a macro-capsule?
4. Which factors need to be taken into account to develop an efficient macro-encapsulated LHTES unit?

In this work, these questions are answered using a systematic approach. The methodology developed to study LHTES units at different scales is structured in the form of different numerical models. Experiments are employed to validate these numerical models at multiple scales. The structure of the present work is depicted as:

- The detailed description of state of the art in chapter 4 draws the motivation for this work.
- The core of work lies in multiscale modelling of LHTES systems, which is described in chapter 5. The distinctive numerical models presented in the chapter 5 illustrate the governing equations required to model the latent heat thermal storage units at different scales.
- Experimental set-up adapted to validate the numerical models is described in chapter 6. The experiments range from studying the solid-liquid phase change in a small scale cavity

to investigating the thermal charging and discharging of laboratory scale thermal storage unit.

- The boundary conditions and material properties of the experiments are adapted into computations with developed numerical models. These computed results from numerical models are compared with experiments in chapter 7. Deviations of numerical results from experiments is also discussed in chapter 7. In chapters 5 and 7, the methodology to compute the charging and discharging of a LHTES unit from a small scale capsule is presented.

- In chapter 8, the influence of capsule material and capsule wall thickness are studied with the developed large scale storage model.

The systematic approach developed to establish an efficient storage configuration is shown in Fig. 2.1.

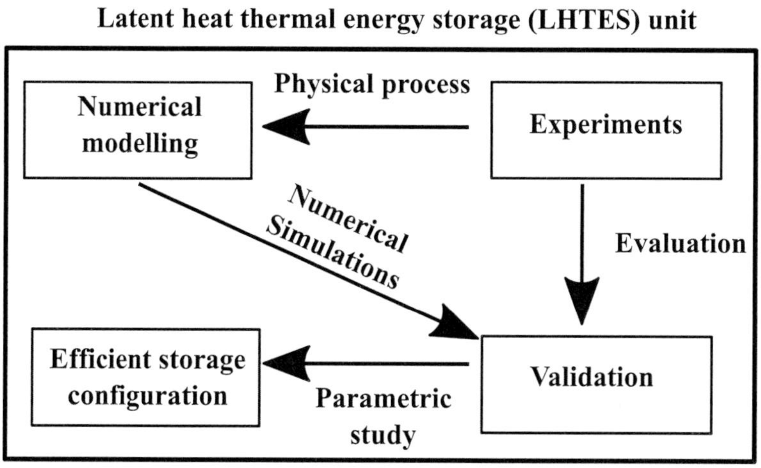

Latent heat thermal energy storage (LHTES) unit

Numerical modelling
Physical process
Experiments
Numerical Simulations
Evaluation
Efficient storage configuration
Parametric study
Validation

Figure 2.1: Overview of steps involved to develop an efficient storage configuration.

3 Computational Thermo-Fluid Dynamics

Physical phenomena involving heat transfer and fluid flow are found in different industrial and natural processes. Scientific understanding of such processes can be improved by studying them in detail. Different physical and operating conditions need to be investigated for a detailed understanding of such processes. Therefore, a conclusive study of any physical process demands a huge investment of time and resources. At the same time it is also difficult to understand complex physical phenomena without having a concrete understanding of each parameter. Numerical modelling in this context may reduce the effort needed to investigate a physical process in detail. Here, a physical process can be represented in the form of numerical model using governing equations. The work-flow involved to find a physical solution for an existing physical process using a numerical model is shown in Fig. 3.1.

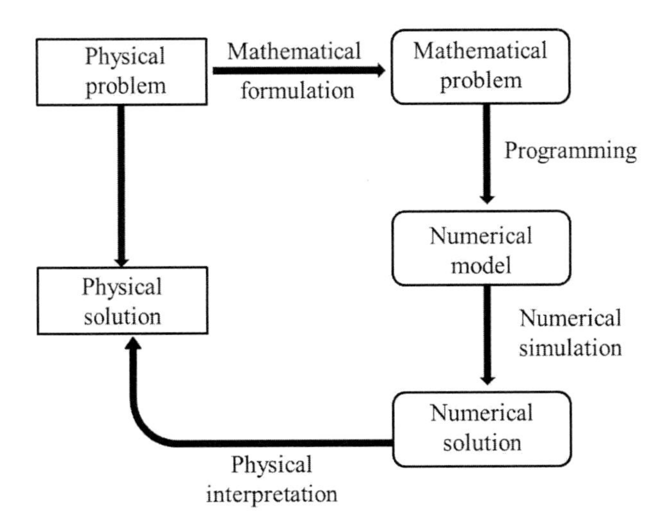

Figure 3.1: Work flow chart depicting the steps to represent a physical process in the form of a numerical model.

The reality is formulated as a mathematical problem, which is implemented in the form of a numerical model using computer programming. A numerical model employs mathematical equations to obtain solutions with defined boundary conditions of a physical problem. In general, a numerical model can be solved to obtain strong or weak solutions, which depends on the mathematical formulation. For a weak formulation, the accuracy of the numerical model depends on its problem definition and solution methodology. In computational thermo-fluid dynamics, physical phenomena are represented in the form of governing equations, which are solved to conserve basic governing laws in a computational domain [6].

Here, equations for the conservation of mass, momentum and scalar are employed [7]. The fundamental governing equations representing the fluid motion are [8]

$$\frac{Dm}{Dt} = 0, \tag{3.1}$$

$$\frac{D\mathbf{M}}{Dt} = \frac{D(m\mathbf{u})}{Dt} = \sum \mathbf{F}. \tag{3.2}$$

Here, m is mass, t is time, $\mathbf{M}$ is the momentum, $\mathbf{u}$ is the velocity vector and $\mathbf{F}$ are forces acting on the mass m. The governing equations represent the time rate change of mass and momentum of a material element, which is derived from the Newton's first and second law of motion respectively [6]. Similarly, the transport of a scalar quantity ϕ can be represented in the form [8]

$$\frac{D\phi}{Dt} = \sum f_\phi. \tag{3.3}$$

Scalar sources and sinks acting on the quantity ϕ is represented by quantity f_ϕ in Eq. 3.3. The scalar quantity ϕ represents physical quantities in chemical and physical processes. Derivative $\frac{D}{Dt}$ in the equations is defined as the material derivative of any physical quantity in continuum mechanics. An elaboration of material derivative for any physical quantity φ reads [7]

$$\frac{D\varphi}{Dt} = \frac{\partial \varphi}{\partial t} + \nabla \cdot (\varphi \mathbf{u}). \tag{3.4}$$

In Eq. 3.4, $\nabla \cdot \mathbf{u}$ is the divergence of velocity, which is the time rate change of moving fluid element volume per unit volume [7]. In Eq. 3.4, the second term on right hand side represents the advection of physical quantity φ, which can be written as

$$\nabla \cdot (\varphi \mathbf{u}) = u_x \frac{\partial \varphi}{\partial x} + u_y \frac{\partial \varphi}{\partial y} + u_z \frac{\partial \varphi}{\partial z}. \tag{3.5}$$

Here u_x, u_y and u_z are velocities corresponding to directions x, y and z respectively. Governing equations in Eq. 3.1, Eq. 3.2 and Eq. 3.3 can be represented in generic form of a transport equation. The general form of transport equation for a physical quantity ϕ reads [9]

$$\frac{\partial \rho \phi}{\partial t} + \nabla \cdot (\rho \mathbf{u} \phi) = \nabla \cdot (\Gamma \nabla \phi) + S. \tag{3.6}$$

The term $\nabla \cdot (\Gamma \nabla \phi)$ is a diffusive term of transport equation. Here, Γ is diffusiveness of the quantity ϕ, ρ is the density and S represents corresponding sources and sinks of transport equation. Newton's third law deals with the conservation of energy in a system [6]. Accordingly, energy conservation in thermo-fluid processes is taken into account in the numerical modelling with a governing equation. Accordingly, Eq. 3.3 and Eq. 3.6 can be used to represent the simple form of energy conservation equation as [8]

$$\frac{D\rho h}{Dt} = \frac{\partial \rho h}{\partial t} + \nabla \cdot (\rho \mathbf{u} h) = \nabla \cdot (\kappa \nabla T) + S. \tag{3.7}$$

Here, h is the internal energy or specific enthalpy, T is temperature and κ is the thermal conductivity. The general transport equation as in Eq. 3.6 can be reformulated to represent the extensive form of mass, momentum and energy conservation [9].

Table 3.1: Specification of the general transport equation [9].

Equation	ϕ	$\nabla \cdot (\Gamma \nabla \phi)$	S
Conservation of mass	1	0	0
Conservation of momentum	$\mathbf{u}$	$\nabla \cdot \tau$	$-\nabla p + \rho \cdot \mathbf{g} + S_{mom}$
Conservation of energy	h	$-\nabla \cdot \mathbf{q}''$	$\frac{\partial p}{\partial t} + \nabla \cdot (\tau \cdot \mathbf{u})$

The general transport equation can be stipulated from Tab. 3.1 to obtain a specific conservative equation. To conserve mass in computational domain, the governing equation is formulated as

$$\frac{\partial \rho}{\partial t} + \nabla \cdot (\rho \mathbf{u}) = 0. \tag{3.8}$$

In similar way, transport equation for the conservation of momentum is formulated as

$$\frac{\partial \rho \mathbf{u}}{\partial t} + \nabla \cdot (\rho \mathbf{u}\mathbf{u}) = \nabla \cdot \tau - \nabla p + \rho \cdot \mathbf{g} + S_{mom}. \tag{3.9}$$

In Eq. 3.9, S_{mom} represents sources and sinks in the momentum equation. Here, p is the pressure, $\mathbf{g}$ is the gravitational acceleration and τ represent the shear stresses in fluid. The shear stresses are derived from Cauchy stress tensor $\mathbf{T}$ used to represent the total stresses acting on a particle [8, 10]. Accordingly, extensive form of equation to represent the conservation of energy reads

$$\frac{\partial \rho h}{\partial t} + \nabla \cdot (\rho \mathbf{u} h) = -\nabla \cdot \mathbf{q}'' + \frac{\partial p}{\partial t} + \nabla \cdot (\tau \cdot \mathbf{u}). \tag{3.10}$$

Here, $\mathbf{q}''$ is the heat flux vector in energy equation. The different representation of energy equation in Eq. 3.7 is a simplified form of the comprehensive energy equation shown in Eq. 3.10. The rate of change of pressure in Eq. 3.10 can be neglected for incompressible flows and also for compressible flows with minimal dependency on pressure change.

The total stress tensor in the form of Cartesian coordinates can be written as [10]

$$\mathbf{T}_{ij} = \begin{bmatrix} \sigma_{ii} & \tau_{ij} & \tau_{ik} \\ \tau_{ji} & \sigma_{jj} & \tau_{jk} \\ \tau_{ki} & \tau_{kj} & \sigma_{kk} \end{bmatrix}. \tag{3.11}$$

Here, σ_{ii} contribute to stresses in normal directions and τ_{ij} are shear stresses of the total stress tensor T. For Newtonian fluids, the stress tensor T is the molecular rate of momentum transport which can be formulated as [8]

$$\mathbf{T}_{ij} = -\left(p + \frac{2}{3}\mu\frac{\partial u_j}{\partial x_j}\right)\delta_{ij} + 2\mu\mathbf{D}_{ij}. \tag{3.12}$$

In Eq. 3.12, μ is the dynamic viscosity, δ_{ij} is Kronecker delta and δ_{ik} is 1 for $i = k$, which contribute to the diagonal elements of matrix and 0 in other cases. D_{ij} in the above equation is the deformation tensor representing the rate of strain as

$$\mathbf{D}_{ij} = \frac{1}{2}\left(\frac{\partial u_i}{\partial x_j} + \frac{\partial u_j}{\partial x_i}\right). \tag{3.13}$$

From Eq. 3.13, the shear stress tensor τ can be represented as

$$\tau_{ij} = 2\mu\mathbf{D}_{ij} - \frac{2}{3}\mu\frac{\partial u_j}{\partial x_j}\delta_{ij}. \tag{3.14}$$

Accordingly from Eq. 3.12 and Eq. 3.14, momentum conservation in Eq. 3.9 can be written as

$$\frac{\partial \rho \mathbf{u}}{\partial t} + \nabla\cdot(\rho\mathbf{u}\mathbf{u}) = -\nabla p + \nabla\cdot(\mu\nabla\mathbf{u}) - \frac{2}{3}\nabla\mu(\nabla\cdot\mathbf{u}) + \rho\cdot\mathbf{g} + S_{mom}. \tag{3.15}$$

Eq. 3.15 is the generic form of momentum equation representing the forces acting on a fluid particle responsible for its rate of change of momentum. The general form of conservative transport equations are represented in Eq. 3.8, Eq. 3.15 and Eq. 3.10.

In fluid mechanics, sometimes the influence of some physical quantities in a process can be neglected. In such case, conservation of these physical quantities can be excluded from the numerical model. For example, in an isothermal process it is less relevant to consider the conservation of energy to represent the physical process. Therefore, in computational thermo-fluid dynamics such physical effects can be represented in simplified forms. Some of such simplifications include incompressible, inviscid, creeping and potential flows [8]. For an isothermal flow with constant density and viscosity, the flow can be considered incompressible. Here, mass and momentum conservation equations are solved to represent the physical phenomena. The corresponding equation for mass conservation from Eq. 3.8 for incompressible flow becomes

$$\nabla\cdot\mathbf{u} = 0 \tag{3.16}$$

due to the constant density. In incompressible flow divergence of velocity is zero, which means that the time rate of change of fluid volume per unit volume is zero. Similarly, the momentum conservation equation for incompressible flow is derived from Eq. 3.15 and Eq. 3.16 resulting in

$$\rho\left[\frac{\partial \mathbf{u}}{\partial t} + \nabla\cdot(\mathbf{u}\mathbf{u})\right] = -\nabla p + \nabla\cdot(\mu\nabla\mathbf{u}) + \rho\cdot\mathbf{g} + S_{mom}. \tag{3.17}$$

Density specific form of momentum conservation equation becomes

$$\frac{\partial \mathbf{u}}{\partial t} + \nabla \cdot (\mathbf{u}\mathbf{u}) = -\frac{\nabla p}{\rho} + \nabla \cdot (\nu \nabla \mathbf{u}) + \mathbf{g} + \frac{S_{mom}}{\rho}. \tag{3.18}$$

Here, ν is the kinematic viscosity of fluid which is the ratio between dynamic viscosity of fluid and its density, $\frac{\mu}{\rho}$. Special cases of incompressible flow with non-isothermal behaviour exist where change in density cannot be neglected. Such fluid flows are associated with heat transfer where physical quantities are dependent of temperature. For such flows, where fluid flow is driven by density change, small density changes can be treated linear with Bousinessq approximation [11]. From Bousinessq approximation, the small density change in thermo-fluid flow can be expressed as

$$(\rho - \rho_0)\mathbf{g} = -\rho_0 \beta (T - T_0)\mathbf{g}, \tag{3.19}$$

where ρ_0 corresponds to the density of fluid at temperature T_0 and β is the thermal expansion coefficient of fluid between temperatures T and T_0. The Bousinessq approximation is generally appropriate for small density changes including linear density change.

3.1 Numerical Discretization and Approximation Schemes

Conservative governing equations enable representation of a physical phenomenon in the form of a numerical model. The numerical model is then solved for a specific problem definition with clearly defined geometry and boundary conditions. The spatial description of problem definition is governed by geometrical dimensions. Material distribution of fluid element in the defined geometry is considered as a continuum. In a continuum, fluid particle in a 3D Euclidean space can be represented as set of particles, which experience motion under the influence of external force[12]. An exact solution of the governing equations on a complete domain cannot be solved at a strech. Therefore, the geometry is discretized into small discrete elements, where the governing equations are solved. For a decent discretization, the discretized form of solution is similar to the solution for complete domain without discretization. Fig. 3.2 depicts spatial discretization of a 1D and 2D geometrical configuration.

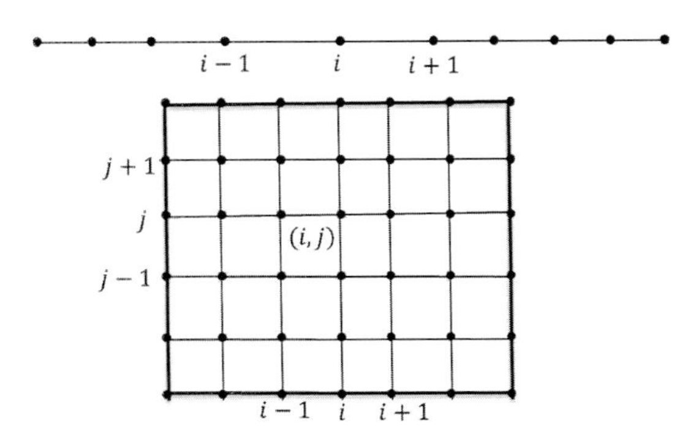

Figure 3.2: Spatial discretization of a 1D and 2D geometry.

In discretized form, the solution is computed at the geometrical nodes. Derivatives and physical quantities of governing equations are computed at each nodal points. Uniform distribution of nodes over a complete domain converge faster due to its better consistency. The distribution of nodal points at equidistant intervals on a computational domain is called structured grid. It is relatively simple to discretize simple geometry into a structured grid, whereas it is difficult for complex geometries. An unstructured grid consists of different cell polygons like polyhedral, tetraherdral and pyramid shape, which converges slower due to uneven aspect ratios.

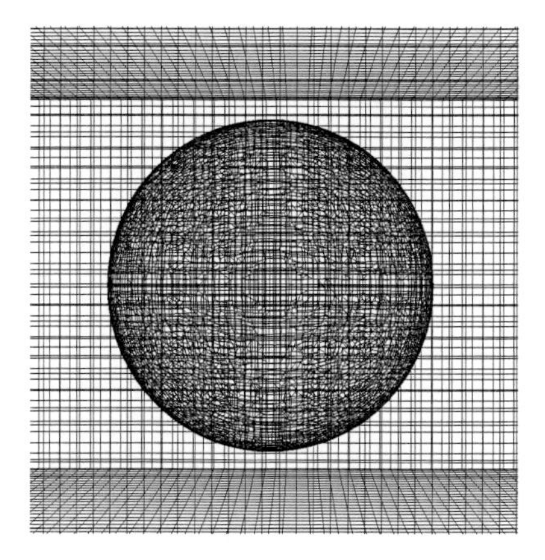

Figure 3.3: Unstructured computational grid of a sphere immersed in a structured grid.

Finite Difference Method

Conservative governing equations of fluid flow and heat transfer are in differential form. In finite difference method (FDM), derivatives in partial differential equation at different computing points are estimated with approximation methods. The expressions of derivative approximations are obtained from basic Taylor series expansion [8]. The Taylor series expansion for derivatives inside a general transport equation Eq. 3.6 for physical quantity ϕ is

$$\phi_{i+1,j} = \phi_{i,j} + \left(\frac{\partial \phi}{\partial x}\right)_{i,j} \Delta x + \left(\frac{\partial^2 \phi}{\partial x^2}\right)_{i,j} \frac{(\Delta x)^2}{2} + \left(\frac{\partial^3 \phi}{\partial x^3}\right)_{i,j} \frac{(\Delta x)^3}{6} + \cdots. \tag{3.20}$$

Initial term on right hand side of Eq. 3.20 is an easy approximation from the previous nodal value. The second term is an approximation of linear slope at nodal point i. The third term is of third order, which induces curvature to approximation. In Eq. 3.20, Δx is grid spacing between the nodal point i and $i+1$. Here, $\phi_{i,j}$ is the value of physical quantity at point i. A similar expansion exists for indices j of 2D geometry and for j and k of 3D geometry. By rearranging Eq. 3.20, the first derivative can be written as

$$\left(\frac{\partial \phi}{\partial x}\right)_{i,j} = \frac{\phi_{i+1,j} - \phi_{i,j}}{\Delta x} - \left(\frac{\partial^2 \phi}{\partial x^2}\right)_{i,j} \frac{(\Delta x)^2}{2} - \left(\frac{\partial^3 \phi}{\partial x^3}\right)_{i,j} \frac{(\Delta x)^3}{6} - \cdots. \tag{3.21}$$

The first order accuracy of derivative can be obtained from first term on the right hand side of Eq. 3.21. Hence, the equation can be written as

$$\left(\frac{\partial \phi}{\partial x}\right)_{i,j} = \frac{\phi_{i+1,j} - \phi_{i,j}}{\Delta x} + O(\Delta x). \tag{3.22}$$

In Eq. 3.22, $O(\Delta x)$ is the truncation error of first-order approximation. Thereby, the first-order approximation of partial derivative is shown in Eq. 3.22. Here, forward-difference approximate is employed for first order approximation. Whereas, the partial derivative approximation using first order backward-difference becomes

$$\left(\frac{\partial \phi}{\partial x}\right)_{i,j} = \frac{\phi_{i,j} - \phi_{i-1,j}}{\Delta x} + O(\Delta x). \tag{3.23}$$

In some cases, first order forward and backward-difference are not relevant to compute an accurate solution. For such, a hybrid central-difference can be used to approximate the derivative. The first order central-differencing scheme is obtained by adding Eq. 3.22 and Eq. 3.23, which reads [7]

$$\left(\frac{\partial \phi}{\partial x}\right)_{i,j} = \frac{\phi_{i+1,j} - \phi_{i-1,j}}{2\Delta x} + O(\Delta x)^2. \tag{3.24}$$

The truncation error in all approximation cases is dependent on the computing cell size Δx. The truncation error increases with the increase in cell size value. In some cases, first order approximation of derivative is not sufficient. In such cases, a higher order approximation can be employed for a better accuracy. Polynomial interpolation of higher order is only necessary if fitting of the curve using linear approximation is not substantial. A parabolic fitting of first derivative yields [8]

$$\left(\frac{\partial \phi}{\partial x}\right)_{i,j} = \frac{\phi_{i+1,j}(\Delta x_i)^2 - \phi_{i-1,j}(\Delta x_{i+1})^2 + \phi_{i,j}\left[(\Delta x_{i+1})^2 - (\Delta x_i)^2\right]}{\Delta x_{i+1}\Delta x_i(\Delta x_i + \Delta x_{i+1})}. \tag{3.25}$$

Substitution of these approximations in the governing equations yields a set of equations for different nodal points. These sets of equations are solved with algebraic matrix approaches. Linearised form of discretized equation at a computing node in algebraic form can written as

$$a_i\phi_i + \sum_{nb} a_{nb}\phi_{nb} = q_i\ , \tag{3.26}$$

where i is the computing node, nb represents the neighbouring nodes, a denotes the coefficients of discretized equation and q_i are terms with independent variables. Similarly, a set of algebraic equations for all computing nodes are obtained. The complete set of algebraic equations are represented in the form

$$\mathbf{A}\phi = \mathbf{Q}. \tag{3.27}$$

In Eq. 3.27, **A** is a matrix with coefficients of discretized equations at all nodes, **Q** is a dimensioned vector carrying all independent variables and sources of discretized equation and ϕ is the variables with the similar matrix dimension as **Q**.

For unsteady problems, the time rate change of a field ϕ exists in the mathematical definition. This is mathematically represented as a derivative of field ϕ with respect to time t, as seen in Eq. 3.6. For such problem definition, the time derivative of field is also approximated. A simple unsteady diffusion problem can be written as

$$\frac{\partial \phi}{\partial t} = \frac{\Gamma}{\rho}\frac{\partial^2 \phi}{\partial x^2}. \tag{3.28}$$

An implicit form of approximation for the Eq. 3.28 can be represented as

$$\phi_i^{m+1} = \phi_i^m + \Delta t \cdot \frac{\Gamma}{\rho} \cdot \frac{\partial^2 \phi_i^{m+1}}{\partial x^2}. \tag{3.29}$$

Whereas, the explicit form is approximated as

$$\phi_i^{m+1} = \phi_i^m + \Delta t \cdot \frac{\Gamma}{\rho} \cdot \frac{\partial^2 \phi_i^{m}}{\partial x^2}. \tag{3.30}$$

Implicit and explicit schemes of the unsteady term are derived from the forward and backward schemes as seen in Eq. 3.22 and Eq. 3.23 respectively. A more accurate method is the hybrid scheme or Crank-Nicolson method [13]. Within Crank-Nicolson method an approximation of the unsteady term reads

$$\phi_i^{m+1} = \phi_i^m + \Delta t \cdot \frac{\Gamma}{2\rho} \cdot \left(\frac{\partial^2 \phi_i^{m+1}}{\partial x^2} + \frac{\partial^2 \phi_i^{m}}{\partial x^2}\right). \tag{3.31}$$

Finite Volume Method

Finite difference method is not substantial to represent physical processes in complex phenomena. For such cases, a finite volume method (FVM) is suitable, which uses a similar discretization concept as FDM. But, here the computational domain is discretized into finite volumes. Solution obtained for a computational domain is similar to the solution obtained by sum of all control volumes. Accordingly, the finite volume integral of general transport equation Eq. 3.6 reads

$$\int_{CV} \frac{\partial \rho \phi}{\partial t}\, dV + \int_{CV} \nabla \cdot (\rho \mathbf{u} \phi)\, dV = \int_{CV} \nabla \cdot (\Gamma \nabla \phi)\, dV + \int_{CV} S\, dV. \tag{3.32}$$

Where CV is the finite volume of the computing cell. According to Gauss divergence theorem the volume integral can be written in a surface integral form [14].

$$\int_{CV} \nabla \cdot (\rho \mathbf{u} \phi)\, dV = \int_{CS} (\rho \mathbf{u} \phi) \cdot \mathbf{n}\, dS. \tag{3.33}$$

Here, CS is the surface area of computing cell CV. If a vector field has continuous first-order partial derivative on computational domain, both sides of Eq. 3.33 have equal meaning [14]. The similar correlation applies for the diffusion term in Eq. 3.32. Accordingly, the general transport equation for a finite volume is written as

$$\int_{CV} \frac{\partial \rho \phi}{\partial t} \, dV + \int_{CS} (\rho \mathbf{u} \phi) \cdot \mathbf{n} \, dS = \int_{CS} (\Gamma \nabla \phi) \cdot \mathbf{n} \, dS + \int_{CV} S \, dV. \tag{3.34}$$

Here, the general form of a transport equation in Eq. 3.34 can be formulated into governing equations for the conservation of mass, momentum and energy. The volume and surface integrals in Eq. 3.34 are approximated with integral approximations. The surface integral of a computing finite volume is approximated as

$$\int_{CS} f \, dS = \sum_{k} \int_{CS_k} f \, dS. \tag{3.35}$$

In Eq. 3.35, f represents the convective component $(\rho \mathbf{u} \phi) \cdot \mathbf{n}$ and diffusive component $(\Gamma \nabla \phi) \cdot \mathbf{n}$ of the general transport equation in Eq. 3.34. k is the number of surfaces bounded around a finite cell volume. The number of faces are four for a 2D computational domain and six for a 3D computational domain. The surface integral approximation method formulates that net flux through control volume boundary as the sum of integrals over control volume surfaces [8]. The surface integral can also be approximated with midpoint, trapezoidal or Simpson's rule.

An approximation of volume integral is necessary to compute the other source terms in governing equations, which is calculated from the nodal point value of control volume P. A simple approximation of volume integral is computed from the product of physical value at nodal point and volume of cell CV, which can be represented as

$$\int_{CV} S \, dV \approx S_P \Delta V. \tag{3.36}$$

Here S_P is the value of source term at point P, which is the centroid of FV, ΔV is the volume of computing cell.

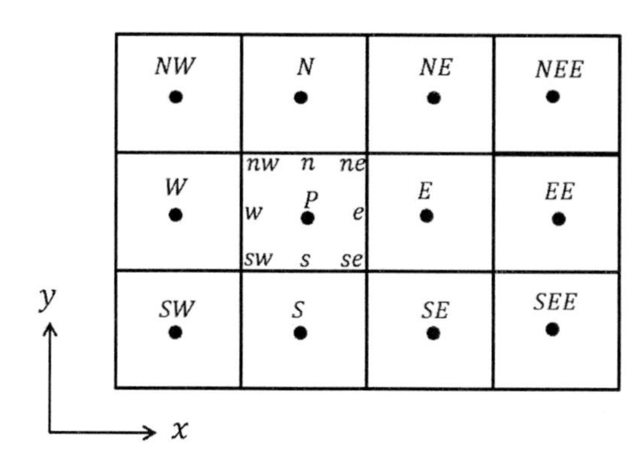

Figure 3.4: Planar view of the computational domain of node P and its respective neighbouring control volumes.

Values at all locations of computational domain are required to approximate the integral correctly. To approximate the convective and diffusive fluxes as in Eq. 3.35, the value of field ϕ and gradient $\nabla\phi$ for a normal vector n need to be known. This is carried out by interpolating the computed nodal value of field ϕ onto cell surface and approximating the gradient $\nabla\phi$ on the cell face. For this, interpolation methods as discussed below are employed.

Upwind Interpolation

Upwind difference scheme or the donor-cell method approximates upstream value of field ϕ_e to nodal value [15]. The approximation employs forward or backward difference schemes depending on the direction of flow [8]. Upwind interpolation approximates the value of field ϕ at eastern cell face e as in Fig. 3.4.

$$\phi_e = \begin{cases} \phi_P, & (\mathbf{u}.\mathbf{n})_e > 0 \\ \phi_E, & (\mathbf{u}.\mathbf{n})_e < 0. \end{cases} \tag{3.37}$$

Where the choice of interpolation on the control volume face is concluded from the product of velocity and normal vector, $(\mathbf{u}.\mathbf{n})$ on the cell face e. The upwind approximation scheme is known for its unconditional stability because of its consistent acceptance of boundedness [8, 16].

Linear Interpolation

Approximation of field value at control volume face as a fraction of distances between two nearest nodes is the fundamental concept of linear interpolation. It is simple and direct approach to compute the field value on the control volume face. The field value on control volume face e reads

$$\phi_e = \phi_E \lambda_e + \phi_P(1-\lambda_e). \tag{3.38}$$

Where λ_e is the fraction factor, which is defined as

$$\lambda_e = \frac{x_e - x_P}{x_E - x_P}. \tag{3.39}$$

Here, x is the coordinate value at position. Using Taylor series expansion, the accuracy of approximation can be improved [8]. The resulting expression for the value at the field on the control volume face e from Eq. 3.20 is

$$\phi_e = \phi_E \lambda_e + \phi_P(1-\lambda_e) - \left(\frac{\partial^2 \phi}{\partial x^2}\right)_P \frac{(x_e - x_P)(x_E - x_e)}{2} - \dots \tag{3.40}$$

In a similar way, gradient at the control volume face can also be approximated. From linear interpolation, the approximate gradient value at control volume face reads

$$(\nabla \phi)_e \approx \frac{\phi_E - \phi_P}{x_E - x_P}. \tag{3.41}$$

Linear interpolation is second-order numerical scheme, which is a popular approach in practice. Due to its calculation of the face at the nodal center, the interpolation is also called the central difference scheme.

Polynomial Interpolation

Higher-order schemes like polynomial interpolation are used to approximate fields only when first and second order schemes deliver a coarse approximation. For such higher-order approximation schemes, values at the control volume face can be fitted using a relevant choice of polynomial coefficients. For uniform grid, the field value on the face e reads

$$\phi_e = \frac{27\phi_P + 27\phi_E - 3\phi_W - 3\phi_{EE}}{48}. \tag{3.42}$$

A gradient on the face can also be approximated using polynomial interpolation. The gradient approximation for a uniform grid yields

$$(\nabla \phi)_e = \frac{27\phi_E - 27\phi_P + \phi_W - \phi_{EE}}{24\Delta x}. \tag{3.43}$$

Interpolation schemes in FVM estimate discretized equation based on the computing node and its neighbours. Therefore, relevant approximation schemes are discussed in the present section of work. Similar to the finite element method, in finite volume method a set of the equation system exist for each computing node. As in Eq. 3.26, even in FVM the system of equations at computing node P reads

$$a_P \phi_P + \sum_{nb} a_{nb} \phi_{nb} = q_P. \tag{3.44}$$

Accordingly, the set of equation system for complete computational domain is represented in a matrix form as

$$\mathbf{A}\phi = \mathbf{Q}. \tag{3.45}$$

Unsteady problems in FVM are solved similar to FDM as shown in Eq. 3.29, Eq. 3.30 and Eq. 3.31. But in FVM, i is defined as the centroid of finite volume P.

Solution for system of equations

Various solving techniques exist to solve matrix equations in the form of Eq. 3.45 and Eq. 3.27. Matrix solution approaches like Gauss elimination, bi-conjugate method, LU decomposition method and multi-grid methods are used to solve the set of equations. The general form of matrix solution for momentum equation reads

$$\mathbf{A}_P\mathbf{u}_P + \sum_{nb} \mathbf{A}_{nb}\mathbf{u}_{nb} = \mathbf{Q}_\mathbf{u} - (\nabla p)_P. \tag{3.46}$$

Here, n refer to neighbouring nodes of point P. Thereby, equation can be formulated as

$$\mathbf{u}_P = \frac{\mathbf{Q}_\mathbf{u} - \sum_{nb} \mathbf{A}_{nb}\mathbf{u}_{nb} - (\nabla p)_P}{\mathbf{A}_P}, \tag{3.47}$$

to obtain the velocity at the computing node P, $\mathbf{u}_P$. Therefore, Eq. 3.47 can be written as

$$\mathbf{u}_P = \mathbf{u}_P^* - \frac{(\nabla p)_P}{\mathbf{A}_P}. \tag{3.48}$$

In Eq. 3.47, velocity at point P where the contribution of pressure gradient can be removed is represented by $\mathbf{u}_P^*$. Nevertheless, at this point the velocity still cannot be calculated due to the unknown influence of pressure gradient. Therefore, continuity equation for incompressible flow is used to estimate this value as

$$\nabla \cdot \mathbf{u}_P = \nabla \cdot \mathbf{u}_P^* - \nabla \cdot \left[\frac{(\nabla p)_P}{\mathbf{A}_P}\right] = 0. \tag{3.49}$$

Accordingly, Eq. 3.49 results in

$$\nabla \cdot \mathbf{u}_P^* = \nabla \cdot \left[\frac{(\nabla p)_P}{\mathbf{A}_P}\right]. \tag{3.50}$$

Semi-implicit method for pressure linked equations (SIMPLE), pressure implicit with splitting of operators (PISO) and some hybrid methods are used to solve the pressure and velocity coupling. The iterations in the algorithm enable strong coupling of velocity with the pressure gradient. A detailed description of these approaches is found in Patankar [17] and Ferziger-Peric [8]. Similar to the momentum equation the energy equation is also reconstructed from the value at the point P. In most cases, the iteration of the energy equation is not necessary unless a non-linear system of equations exist.

3.2 Numerical stability and dimensionless numbers

Steps involved to solve the set of governing equations was discussed earlier in this work. Solution procedures vary depending on the problem definition and set-up. Different numerical parameters like discretization, residual control and step size play a major role in solution process. An unconstrained employment of such parameters or strong jump in these values may lead to oscillating solutions, which deviate from reality. Therefore, some empirical correlations exist to reduce these instabilities in a computational process. A well known convergence criterion for solving partial differential equations is the Courant-Friedrichs-Lewy (CFL) condition [18]. The CFL condition recommends a numerical constraint for computing explicit time marching of a time dependent problems. For one-dimensional case, the condition reads

$$\mathrm{Co} = \frac{u\Delta t}{\Delta x} \leq \mathrm{Co}_{\mathrm{max}}. \tag{3.51}$$

Here, Co is a dimensionless Courant number, u is the velocity magnitude and Δx is the length of cell, Δt is the time step and $\mathrm{Co}_{\mathrm{max}}$ is the maximum value of Courant number. For an explicit time-marching, the CFL condition recommends a maximum value $\mathrm{Co}_{\mathrm{max}} = 1$. Whereas, for an implicit time-marching a value greater than 1 can also be used. Sudden oscillations in explicit time marching can be damped by decreasing the value of $\mathrm{Co}_{\mathrm{max}}$.
Numerical stability of an algorithm depends on the convergence and accuracy. For non-linear problems, numerical stability also lies on the choice of approximation functions and solution methodology. Non-linear partial differential equations like the Navier-Stokes equation have a poor numerical stability. Neumann stability analysis is a possibility to check numerical stability of linear partial different equations. For a simple unsteady diffusion problem as

$$\frac{\partial \phi}{\partial t} = \frac{\Gamma}{\rho}\frac{\partial^2 \phi}{\partial x^2}, \tag{3.52}$$

the Neumann stability is attained when the choice of time step

$$\Delta t < \frac{\rho(\Delta x)^2}{2\Gamma} \tag{3.53}$$

In case of a convection-diffusion problem, where the diffusion dominates the CFL condition

$$\frac{\rho u \Delta x}{\Gamma} < 2 \tag{3.54}$$

ensures the stability of algorithm. Another method to estimate the influence of diffusion is through Péclet number (Pe), which is the ratio between advective and diffusive transport rates :

$$\mathrm{Pe} = \frac{lu}{\Gamma} \tag{3.55}$$

Diffusion coefficient Γ is equal to the mass diffusion coefficient D_m for mass transfer and to thermal diffusivity a for heat transfer, l is the characteristic length. Similarly, Fourier number (Fo) characterizes the heat conduction in a thermal transport process, which is defined as

$$\mathrm{Fo} = \frac{at}{l^2}. \tag{3.56}$$

For advective-diffusive heat transfer problem, heat transfer at the boundary within a fluid is characterized by dimensionless Nusselt number (Nu). This is a ratio between convective and conductive heat transfer at the boundary written as

$$\mathrm{Nu} = \frac{hl}{\kappa}. \tag{3.57}$$

Besides Nusselt number, heat transfer can also be characterized by a dimensionless Rayleigh Number (Ra), which is a ratio between buoyancy and viscosity in a free convecting fluid. For a free convection near a vertical wall, Rayleigh number is defined as

$$\mathrm{Ra} = \frac{\mathrm{g}\beta}{\nu a}\left(T_w - T_\infty\right) l^3. \tag{3.58}$$

Here, ν is the kinematic viscosity of fluid, β is the thermal expansion coefficient, T_w is temperature on the vertical wall surface, T_∞ is the quiescent temperature. Rayleigh number Ra can also be written as the product of Grashof and Prandtl numbers, where Grashof number for a free convection near a vertical wall can be written as

$$\mathrm{Gr} = \frac{\mathrm{g}\beta}{\nu^2}\left(T_w - T_\infty\right) l^3, \tag{3.59}$$

whereas, Prandtl number (Pr) is the ratio of momentum diffusivity and thermal diffusivity written as

$$\mathrm{Pr} = \frac{c\mu}{\kappa}. \tag{3.60}$$

Here, c is the specific heat. In thermodynamics, Stefan number (Ste) indicates the latent heat stored in fluid phase of a material, which is written as

$$\mathrm{Ste} = \frac{c(T_h - T_m)}{L}. \tag{3.61}$$

Where T_m is melting temperature of the substance and T_h is constant temperature of the heat source. In fluid mechanics Reynolds number (Re) describes the nature of flow, defined as a ratio between the inertial and viscous forces, written as

$$\mathrm{Re} = \frac{\rho u l}{\mu}. \tag{3.62}$$

3.3 OpenFOAM

OpenFOAM (Open Field Operation and Manipulation) is an open-source C++ library, which employs finite volume method to solve governing equations for fluid flow and heat transfer. OpenFOAM was initially developed as an in-house code in Imperial College London by Henry Weller group. Since 2007, this C++ library is named OpenFOAM and released open source for public use. A general momentum equation for incompressible flow reads

$$\frac{\partial \rho \mathbf{u}}{\partial t} + \nabla \cdot (\rho \mathbf{u}\mathbf{u}) - \nabla \cdot \mu \nabla \mathbf{u} = -\nabla p. \tag{3.63}$$

For example Eq. 3.63 is formulated in OpenFOAM as

```
solve
(
        fvm::ddt(rho,U)
      + fvm::div(phi,U)
      - fvm::laplacian(mu,U)
    ==
      - fvc::grad(p)
);
```

where phi in OpenFOAM represents the interpolated product of velocity vector $\mathbf{u}$ and density ρ on a finite volume surface. The derivative of a field with respect to time is represent by ddt in OpenFOAM, div is the divergence of a field and grad is the gradient. OpenFOAM employs a three dimensional spatial domain to represent a physical processes [19].

4 State of the art

Existing state of the art methods to compute physical phenomena like phase change, multiphase flows and overlapping grid techniques are discussed in this section of work. A brief description of these methods provide a deep insight of scientific evolution from empirical methods to the present work. This work is inspired from different empirical approaches to bridge gaps in the field of numerical modelling.

4.1 Phase change problems

The transition between one state of matter to another is described as phase change. Phase change is a commonly observed process in different industrial and domestic applications. Transition of phase occurs due to the change in pressure or temperature of a system. Which is driven by the change in molecular structure of matter existing in a phase. Phase change diagram in Fig. 4.1 represents the transition of phase with respect to change in pressure and temperature.

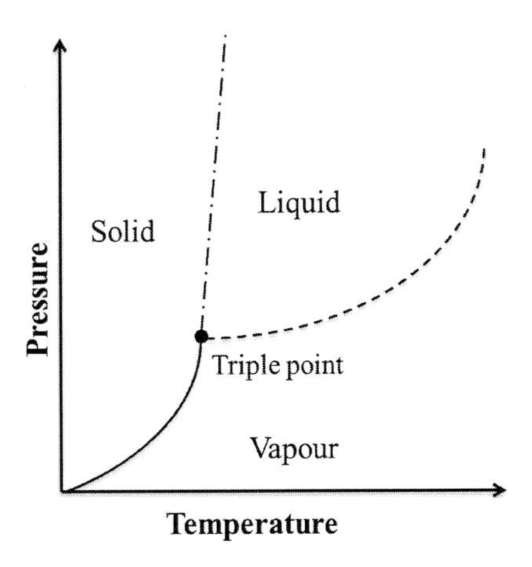

Figure 4.1: Phase diagram representing three states of matter solid, liquid and vapour.

Three lines representing the change of phase between solid-liquid, liquid-vapour and vapour-solid meet at triple point. At triple point solid, liquid and vapour phases of a substance co-exist in equilibrium. In general, phase transition can be studied through various experimental techniques. But, study of phase change with repetitive experiments is very expensive and time consuming. Therefore, numerical modelling and simulation is a cost-effective way to study phase change in different processes. Thus, numerical modelling and simulation of solid-liquid phase change is the topic of interest here. As seen in the Fig. 4.1, solid-liquid phase change occurs due to the increase in internal temperature of a substance, which is occurred through energy consumption from external source. As seen in Fig. 4.2, temperature change with the change in internal energy or enthalpy is non-linear in a solid-liquid phase change process.

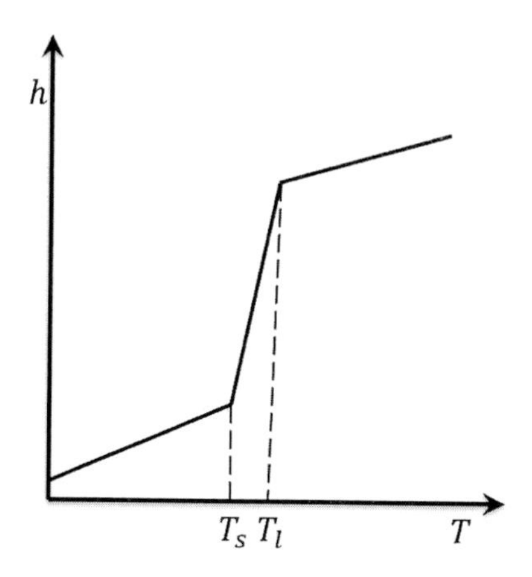

Figure 4.2: Plot depicting non-linear relationship between enthalpy and temperature.

It is challenging to represent such non-linear relationships mathematically. But, Josef Stefan [20] around 1890 initially introduced a mathematical representation of solid-liquid phase change with the help of a Fourier equation. Later, an unsteady diffusion for a moving phase change boundary was mathematically interpreted by Danckwerts [21]. Following this, Crank [22] introduced two different methods to compute heat transfer and diffusion of a moving boundary. Later, Furzeland [23], Crowley [24] and others have employed different modelling approaches to compute the movement of phase change boundary driven by diffusion. Comini et al. [25, 26] implemented the non-linear temperature model for phase change problems in finite element method (FEM). In this work, the latent heat term was not taken into account, instead a variable specific heat was considered. Therefore, Morgan et al. [27] modified this approach to model phase change with an apparent heat capacity. Previously, Shamsundar and Sparrow [28] also introduced the enthalpy model for phase change with multidimensional distinct phase change temperatures. But, in all these numerical models only the diffusion of moving boundary during phase change was implemented. Modelling diffusion alone is not substantial to understand the detailed behaviour of solid-liquid phase change. Therefore, Gartling [29] presented a work on modelling convective heat transfer during solid-liquid phase change. In this work he employed FEM to solve a set of mass, momentum and energy equations. He extended this numerical model to compute the standard Stefan problem along with a solidification casting problem. In this work, Gartling [29] introduced an approach to suppress the solid freedom with a very high viscosity value in the solid domain. The viscosity in this case is a function of temperature meaning $\mu = \mu(T)$, where μ takes a very high value for lower temperatures i.e. in solid phase. Later, Morgan [30] extended his previous model [27] to compute freezing and melting of a substance with convection. Here, Morgan [30] also introduced an approach where the velocity in the solid phase is overwritten by zero. Here, the non-linear temperature-enthalpy plot is represented by an apparent heat capacity taking a peak value in the melting range. The apparent heat capacity proposed by Morgan [30] reads

$$c^A = c^{sen} + [H(T - T_s) - H(T - T_l)] \frac{L}{T_l - T_s} \tag{4.1}$$

where c^{sen} is the specific heat capacity of sensible part, c^A is the apparent heat capacity, H is a Heaviside step function, T_s is solidus temperature where the substance stays fully solid, T_l is liquidus temperature where the substance becomes fully liquid and L is the latent heat released or absorbed during solid-liquid phase change. Despite the contributions from Gartling [29] and Morgan [30], it was not able for researchers [31–33] to employ these approaches for complex applications. During this period, Crank [34] also published a detailed overview on the numerical approaches for conductive solid-liquid phase change. A lumped capacitance representation of the apparent heat capacity was presented by Pham [35], which also ignores convection during phase change. Therefore, Voller et al. [36] including Voller and Prakash [37] proposed a different enthalpy method for fixed grids to compute solid-liquid phase change with heat transfer and fluid flow. A minor difference between these numerical models is the inclusion of latent heat in the convection term of energy equation. Voller and Prakash [37] took the convection of latent heat term into account. Whereas, Voller et al. [36] considered just a time rate change of the latent heat as the source term in energy equation. In addition to the introduction of convective solid-liquid phase change, the work also introduced a porosity source term to ramp down the velocity in solid phase. They introduced a Kozeny-Carman source term in momentum equation to enable fixing of solid phase during solid-liquid phase change. This source term reflects the porosity approach derived from Darcy law. Therefore, in later years it is well known as Darcy source term to model solid-liquid phase change. Following Voller and Prakash [37], the Carman-Kozeny equation following Darcy law reads

$$\nabla p = -C \frac{(1-\gamma)^2}{\gamma^3 + \varepsilon} \cdot \mathbf{u}, \tag{4.2}$$

where C is the Darcy constant, γ is the porosity and ε is a small value restricting the division of numerator by zero. Beckermann and Viskanta [38] adopted a similar approach to represent the solid-liquid phase change with volume averaged transport equations. Here, the non-linear resistances constituting the micro-structure interactions during solid-liquid phase change are taken into account. The results obtained from the numerical model were validated with experiments, which have shown good acceptance [38]. In the same year, Brent et al. [39] also proposed the widely known enthalpy-porosity approach to model solid-liquid phase change. A full implicit coupling of the latent heat source term in the energy equation is presented in this work. Enthalpy-porosity method is an iterative approach enabling the strong implicit implementation of latent heat source term in energy equation. The source term in energy equation contributing to latent heat is defined as

$$S_L = \frac{\partial(\rho h_L)}{\partial t} + \nabla \cdot (\rho \mathbf{u} h_L) \tag{4.3}$$

where S_L is the latent heat source term and h_L is the latent heat enthalpy, which takes a value zero in the solid phase and latent heat value in the liquid phase. Unlike the previous works here a numerical model for pure metals with isothermal phase change was introduced. In this work, the numerical model was also validated with experimental results for melting of gallium by Gau and Viskanta [40]. On the other side, Bennon and Incropera [41] developed a numerical model to represent solid-liquid phase change for impure binary alloys. They introduced a mixture enthalpy model to illustrate the binary mixture advection during phase change. Lewis and Roberts [42] introduced an approach to compute apparent heat capacity from a mean gradient of computing cell, which reads

$$c^A = \frac{dh}{dT} = \left(\frac{\nabla h \cdot \nabla h}{\nabla T \cdot \nabla T}\right)^{0.5}. \tag{4.4}$$

Later, for unsteady phase change problems Dantzig [43] formulated apparent heat capacity as a ratio of time rate change between enthalpy and temperature, which reads

$$c^A = \frac{dh}{dT} = \frac{dh/dt}{dT/dt}. \tag{4.5}$$

Later, Cao et al. [44] represented the phase change numerically as a piecewise linear distribution of the apparent heat capacity. They presented a three-dimensional model for phase change with convection. A more correct implementation of this approach with a combination of source based method is presented by Cao and Faghri [45]. The numerical results obtained in this work have shown a very good acceptance with the experiments from Okada [46]. Due to its simplicity, fixed grid approaches were employed often by different researchers, whereas the moving grid approaches have been investigated only by very few. Lacroix and Voller [47] dealt moving grids alongside providing a comparison between fixed and transforming grid approaches. The numerical results from both approaches were validated with experiments from Gau and Viskanta [40]. Despite a very good acceptance with experiments, authors have reported difficulties in dealing with transforming grids for complex applications. In practice, it is difficult to employ pure materials for solid-liquid phase change. Materials with some impurities are more affordable, but the phase change in them is non-isothermal, which occurs over a temperature range. Accordingly, a phase transition zone or mushy zone is assumed over this temperature range. Therefore, Voller et al. [48] introduced an approach to implement solid particle suspension and liquid mixture in the mushy zone. Later, Prakash and Voller [49] improved the mixture model for accurate and better tracking of mushy zone. Despite its accuracy, this model is not suitable for large scale applications due to its computational expense. Therefore, Swaminathan and Voller [50] came up with an optimized form of enthalpy method. This numerical approach targeted to reduce computational effort by modelling piecewise slope of the non-linear enthalpy and temperature relation. The slope of non-linear enthalpy and temperature relation is the apparent heat capacity c^A, which can be shown in Eq. 4.1. In a later work from Swaminathan and Voller [51], the numerical approach was compared with other numerical methods to reflect the efficiency of this method. For different phase change profiles, the numerical approach was more efficient than the source based method and the apparent heat capacity method. Here, for a convection-diffusion problem discontinuities in thermal conductivity are treated with Kirchhoff transformation [52]. Later, Viswanath and Jaluria [53] also compared the enthalpy method with transforming grid approaches. They reported a very good agreement of numerical solutions with experiments. In general, iterative approaches have shown better results than the non-iterative approaches in computing phase change problems [54]. A comprehensive comparison of different approaches for solving conductive solid-liquid phase change was presented by Pham [55]. In this comparison, the optimized iterative approach from Swaminathan and Voller [50] achieved a very good accuracy within a commendable computational time. Here, the comparison was made between iterative and non-iterative approaches. The optimum approach needed a little more computational time than the other non-iterative approaches but has ensured good heat balance at all time steps. A thorough review of all the existing methods for solid-phase change has provided a clear insight, where the optimum approach has been proven better than apparent heat capacity in its accuracy [56]. Some important contributions include the detailed study of phase change problems in later years [57].

Close contact melting

In general, for most of the materials, the solid phase is denser than its liquid phase. However, some exceptions exist for materials like water. During melting, the denser solid settles to the bottom of geometry resulting in a small melting gape. Unfixed solid moving towards the bottom of geometry is termed as close contact melting [58]. The movement of unfixed solid during melting process was initially investigated by Bareiss and Beer [59].

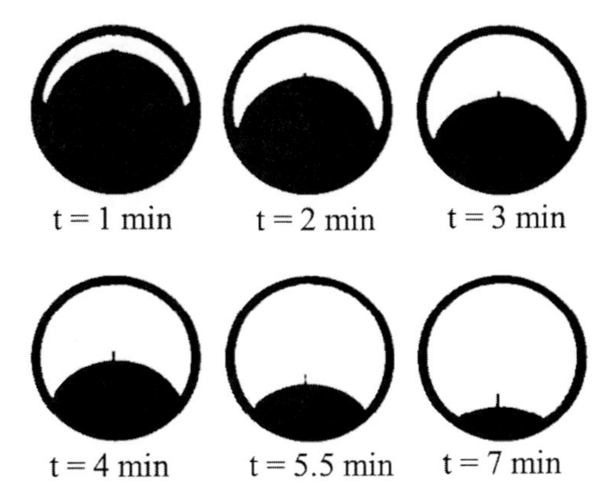

Figure 4.3: Melting process of n-octadecane in a horizontal cylindrical tube [59].

Moallemi et al. [60, 61] also contributed to understand close contact melting in a phase change process. Close contact melting accelerates the melting process increasing the heat flow from geometry bottom [62]. In early years, a simple theoretical correlation was introduced to describe close contact melting [63, 64]. This is however not substantial to understand complex melting and settling of solid phase. Therefore, a detailed numerical model for melting and settling of solid in a cavity was developed [65]. Here, Asako et al. [65] modelled the unfixed solid in a rectangular cavity from the force balance on solid domain. Settling velocity was computed using Newton-Raphson method from the resultant force acting on the solid body. This numerical model was later extended for low gravity and variable gravity environments [66–68].

Later, a relatively simple implementation of the solid settling during melting was introduced by Ghasemi and Molki [69]. Here, settling velocity of solid dependent on the Rayleigh number was studied. Some other authors [70–73] tried to model close contact melting by relating solid settling with dimensionless numbers. Later, the work from Assis et al. [74] to model melting of unfixed solid in a spherical shell has drawn a decent amount of attention. Here, enthalpy-porosity approach was used to model the melting and settling of solid phase in a spherical shell. In this work, the close contact melting is represented in a non physical manner. Tan et al. [75] used polar coordinates instead of Cartesian coordinates to represent the melting of solid in a spherical shell. However, the movement of the solid here is restricted due to the fixing of solid to a thermocouple. Shmueli et al. [76] investigated the melting of paraffin in a vertical cylinder tube using numerical modelling and experiments. In this work, the enthalpy-porosity approach with body-weight settling of the solid was employed. Rösler and Brüggemann [77] later employed an error function to model the melting and settling of paraffin in a shell and tube type latent heat thermal storage system. Also, Hosseinizadeh et al. [78] investigated the melting of n-octadecane in a spherical shell numerically and experimentally. In this work, enthalpy-porosity approach has been used to computed the solid-liquid phase change. A large contribution in this field arouse from Rösler [79], where a more appropriate model for melting of paraffin with enthalpy porosity approach was presented. Solid settling in the work was modelled from force balance on the solid domain. Here, the steady state solid velocity at each time step is computed from explicit treatment of Newton-Raphson method [79]. In this work, a multiphase volume of fluid (VOF) approach was employed to represent the multi-layered fluid system inside a PCM macro-capsule. A simple convective model for close contact melting in a vertical annular enclosure was introduced by Kozak et al. [80], the model solves a simple radial problem definition which with a non-iterative approach. Recently, the authors [81] extended this numerical model to account the force balance on the solid domain. Instead of a conventional Darcy term a discrete Dirac delta function has been used. The variable viscosity method (VVM) employed in the present work is partly contributed from recent works [82, 83].

4.2 Multiphase flows

Flows consisting of multiple substances like impurities, dispersions or inclusions exist in different natural and industrial applications. Each constituent in the flow is referred as a phase. Multiphase flows demand complex understanding of phase interactions including the behaviour of mixture. Due to complex physics, understanding multiphase flows are more intricate than single phase flows. Numerical modelling of multiphase flows improves the understanding of flow. Computation of multiphase flows is a huge field, which exist for the past 40 years. Advancement of modelling techniques in the field of multiphase flows is intensive due to their vast application in diverse fields. Employment of simple numerical techniques to represent the multiphase phenomena was developed very early [84], but a detailed numerical representation of multiphase flows was only achieved after the development of marker and cell method by Harlow [85]. Enormous amount of effort has been invested to develop computer algorithms to represent interfaces between fluids, multiphases and droplets [86–89]. Accordingly, the marker and cell method has been adapted for different applications. Hirt et al. [90] extended the Eulerian representation of marker and cell method by coupling it with Lagrangian method for computing the free surface. Morrison [91] used a volume fraction approach to model the air and mixture vapour in porous media. An important contributed is the modified maker and cell method by Vicelli [92], who introduced tracking arbitrary boundary in incompressible flow. Here, a relative pressure distribution for the multiphase distribution is developed.

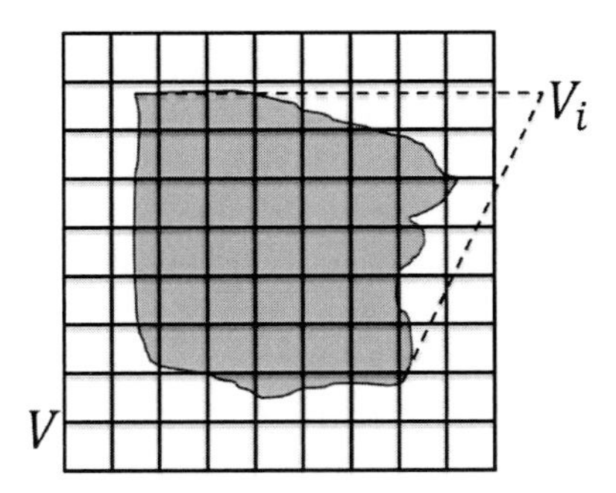

Figure 4.4: Dispersed phase with volume V_i sharing an irregular interface with continuous medium in a computational grid with volume V.

Hirt et al. [93] extended the numerical model with arbitrary Eulerian method to an arbitrary Largangian representation of marker and cell method. Simple Line Interface Calculation (SLIC) by Noh and Woodward [94] has drawn a lot of attention because of its smooth representation of interface between two different states of matter. Crowe and Stock [95] employed tank and tube approach which is later renamed as the FVM to model the movement of particle in continuous fluid flow. A prominent contribution in the field of multiphase modelling is the volume of fluid (VOF) method by Hirt and Nichols [96]. In their work a time dependent volume of fluid equation has been introduced to track the advection of fluid volume in a computational domain. The VOF method is a volume average technique to compute mean distribution of the dispersed phases inside fluid continuum. The heterogeneous distribution of dispersed phase in fluid continuum is homogenized by an averaging technique. Volume averaging of dispersed phase can be represented as

$$\langle \alpha_i \rangle_V = \lim_{V \to \infty} \frac{1}{V} \int_V \alpha_i dV. \tag{4.6}$$

Accordingly, governing equations are adapted taking the movement of multiphase and free surface tracking into account. The VOF equation proposed to represent the continuity of a phase for 2D domain by Hirt and Nichols [96] reads

$$\frac{\partial \alpha}{\partial t} + \mathbf{u}_x \frac{\partial \alpha}{\partial x} + \mathbf{u}_y \frac{\partial \alpha}{\partial y} = 0, \tag{4.7}$$

where α is the volume fraction of fluid phase. In Eq. 4.7, the fluid volume crossing the cell boundary in a single time step Δt per unit cross sectional area is equal to $\mathbf{u}\Delta t$ [96]. The total flux ϕ in a cell for a multiphase flow is calculated using volume averaging approach. Thereby, the total flux ϕ in a cell reads

$$\phi = \sum_{i=1}^{n} \alpha_i \phi_i \tag{4.8}$$

where α_i represents the volume fraction of each phase in a computing cell and ϕ_i its corresponding flux. Following this work, inter-phase-slip algorithm (ISPA) from Spalding [97] is also a pivotal contribution to model fluid flow and heat transfer in multiphase flow. Durst et al. [98] in later years introduced a Eulerian-Lagrangian model for VOF method to represent trajectory of particles in a fluid continuum. The dispersed phase in a continuous flow medium may range between small and large volume fractions. A classification for the range of suspensions in a fluid continuum was illustrated by Elghobashi [99]. Here, the interaction between the fluid and solid particles are characterized based on their influence on the flow domain. For a very low fraction of dispersed phase, one way coupling meaning just the influence of continuous phase on the disperse phase is considered. For higher volume fractions of the disperse phase, the influence of continuous phase on disperse phase along with the influence of disperse phase on continuous phase are considered. For very high volume fractions, the interaction between the particles are also considered as shown in Fig. 4.5.

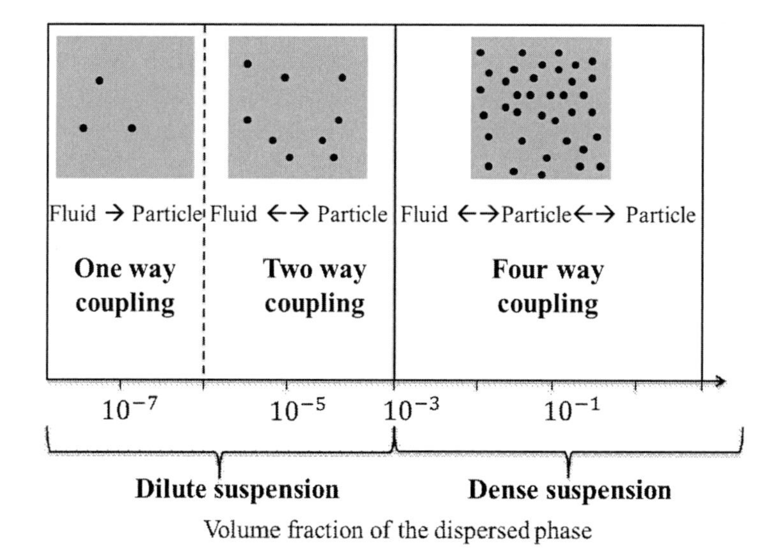

Figure 4.5: Classification of coupling methods between dispersed and continuous phase [99].

In general, the volume fraction of a phase is defined as the

$$\alpha_i = \frac{V_i}{V}, \tag{4.9}$$

where α_i is the volume fraction of the phase i, V_i represents the volume of phase i and V represents the total domain volume as shown in Fig. 4.6.

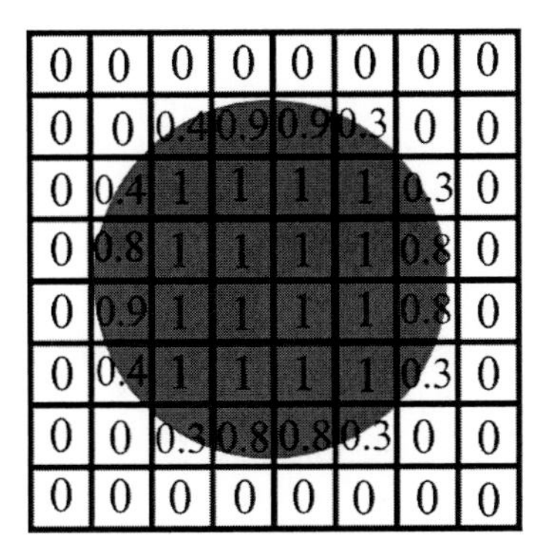

Figure 4.6: Volume fraction distribution of spherical disperse phase in a computational grid.

Some other studies [100–102] introduced various approaches to represent multiphase flows in different applications. Rider and Kother [103] developed an algorithm to reconstruct dispersed phase volume and the interface. In their work, a VOF method was employed to track the interface between dispersed and continuous phase. Later, Sussman et al. [104, 105] developed a level set approach to compute interfacial flows. Here, a level set function φ is employed to represent multiphase phenomenon. The level set function has a value of 0 at the interface and +1 and -1 in each phase respectively. A heaviside step function is employed to model the surface tension between two different phases are modelled in this work. For this, a surface tension model similar to the continuum surface force (CSF) model from Brackbill et al. [106] is employed. The volumetric formulation of surface tension force proposed by proposed by Brackbill et al. [106] is

$$\mathbf{F}_{sf} = \sigma \kappa_{sf} \cdot \nabla \alpha, \tag{4.10}$$

where σ is the surface tension coefficient and κ_{sf} is the curvature and is defined as

$$\kappa_{sf} = -\nabla \cdot \mathbf{n}. \tag{4.11}$$

Here, $\mathbf{n}$ is the unit normal vector. Surface tension force at the interface from Eq. 4.10 is implemented in momentum conservation equation. Despite a very sharp tracking of the interface, the level set method may deliver non satisfactory results due to its poor mass conservation. Therefore, Sussman and Puckett [107] came up with a coupled level set and volume of fluid (CLSVOF) method. CLSVOF method combines both volume of fluid (VOF) and level set (LS) method for better accuracy. In this method, the poor mass conservation in LS method is countered by the VOF equation as in Eq. 4.7. Numerical diffusion at the interface with VOF method is optimized by the level set function in LS method. Torres and Brackbill [108] introduced a point-set method using Dirac delta function for a sharp tracking of the interface. A hybrid particle level set method from Enright et al. [109] solves an additional signum function to reconstruct the marker particles, which counters the non-conservation of mass from LS method. Later, Rusche [110] developed a detailed numerical model to solve dispersed phase with higher volume fractions in FVM. In this work, inter-facial surface tension force, momentum transfer between the phases and relative velocity at the interface were additionally included to represent multiphase flows. Within this work, a new term for interfacial compression is introduced in the VOF equation. The corrected VOF equation to model phase continuity is written as [110, 111]

$$\frac{\partial \alpha}{\partial t} + \nabla \cdot (\alpha \mathbf{u}) - \nabla \cdot (\alpha(1-\alpha)\mathbf{u}_r) = 0. \tag{4.12}$$

Here, $\mathbf{u}_r$ is the relative compression velocity to compensate different physical effects on either sides of the interface.The last term on the LHS in Eq. 4.12 contributes only to the thin interface between phases where relative velocity is imposed. Relative compression velocity induces an artificial compression of interface between the phases which is denoted as

$$\mathbf{u}_r = \mathbf{n} \min \left[c_\alpha \frac{|\phi|}{|S_f|}, \max \left(\frac{|\phi|}{|S_f|} \right) \right], \tag{4.13}$$

where ϕ is the mass flux, S_f is the cell surface area, c_α is the compression coefficient, which take scalar values. Value for c_α = 0 represents no compression of the velocity, whereas 1 corresponds to a conservative compression. For an enhanced compression, a value greater than 1 can be considered. Later, Shin et al. [112] developed an accurate representation of surface tension by employing a reconstruction method of color function. A comparative study between the CLSVOF, parabolic reconstruction of surface tension (PROST) and K_8 kernel approach was executed by Gerlach et al. [113], to study the performance of these models. PROST has proven to be a better surface tension model followed by CLSVOF and K_8 kernel approach. The surface tension at the interface between two different fluids or states of matter influence the nature of interface between them. The resultant surface force acting on the interface forms a contact angle. The nature of capillarity and meniscus between two mediums is influenced by contact angle. Vertical and horizontal forces at contact point on the surface forms an equilibrium [114]. The process where the liquid maintains a contact with the surface is called wetting. The force balance at the contact point influence the nature of wetting i.e. either adhesive or cohesive. The contact angle defines the nature of wetting of the liquid meaning a small contact angle ($<$ 90°) leads to a high wetting whereas a large contact angle ($>$ 90°) corresponds to a low wetting [115].

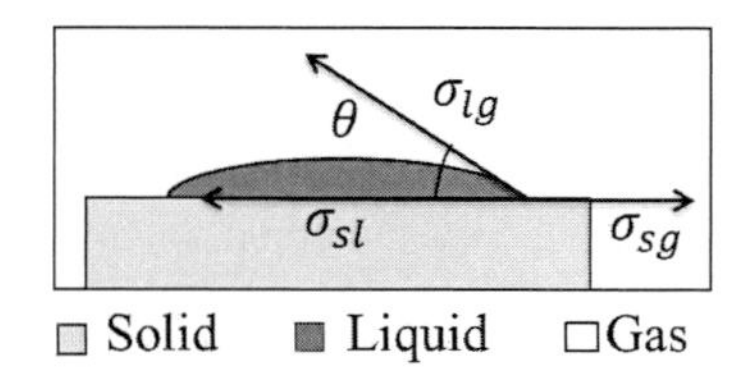

Figure 4.7: Surface tension forces acting over a droplet spread on a solid surface with a contact angle θ.

The contact angle at the surface is measured at the contact point [116]. The force equilibrium obtained from resultant surface tensions reads

$$\sigma_{lg} \cdot \cos\theta = \sigma_{sg} - \sigma_{sl} \tag{4.14}$$

whereby, the contact angle is obtained from

$$\theta = \cos^{-1}\left[\frac{\sigma_{sg} - \sigma_{sl}}{\sigma_{lg}}\right]. \tag{4.15}$$

Here, σ_{lg} is the surface tension between liquid and gas phases, σ_{sg} is the surface tension between gas and solid surfaces, σ_{sl} is the surface tension between solid and liquid surfaces and θ is the contact angle between liquid-gas interface and solid surface as shown in Fig. 4.7. In a numerical model, contact angle is taken into account to realize the interaction between inter-facial surface tension of fluid and wetting boundary. Young-Laplace equation [117] defines surface tension as a correlation of pressure difference across the interface, which reads

$$\Delta p = -\sigma \nabla \cdot \mathbf{n}, \tag{4.16}$$

where Δp is pressure difference across the interface and σ is the corresponding surface tension. Shikhmurzaev [118] and Saha-Mitra [119] proposed a dynamic contact angle which is a function of the capillary number (Ca). The dynamic contact angle proposed by Shikhmurzaev [118] reads

$$\cos\theta_d = \cos\theta_e - \frac{2u(a_1 + a_2 u_0)}{(1-a_2)[(a_1+u^2)^{0.5}+u]} \tag{4.17}$$

where θ_d is the dynamic contact angle, θ_e is the static contact angle, a_1, a_2, a_3 and a_4 are the phenomenological constants. Saha and Mitra [119] additionally performed a comparative study on different dynamic contact angle models. Accordingly, they have employed dynamic contact angle relations from Kalliadasis and Chang [120], Sikalo et al. [121] and Bracke [122]. The effect of dynamic contact angle in a microfluidic capillary flow was investigated here. The results obtained from the numerical simulations were validated with experiments. Here, the effect of dynamic contact angle has shown a very minimal influence. Wang et al. [123] implemented a full CLSVOF method to model sharp interfaces by interface reconstruction using piecewise linear interface construction (PLIC) scheme. The numerical model has shown a very good prediction of the reality in different set-up conditions. A similar numerical model to calculate bubble growth and detachment was presented by Albadawi et al. [124]. This numerical model was implemented in OpenFOAM and a static contact angle was employed. In their work, the detachment and evolution of a droplet from a surface to a fluid field is computed and validated with experiments. Similarly, Malgarinos et al. [125] investigated the influence of dynamic contact angle on droplet wetting. Dynamic contact angle in OpenFOAM is defined as a function of advancing angle, receding angle and velocity at the wall. The dynamic contact angle implemented in OpenFOAM reads [126]

$$\theta^n = \theta^{n-1} + (\theta_A - \theta_R) \cdot \tanh\left[\frac{\mathbf{u}_{wall}}{\mathbf{u}_\theta}\right]. \tag{4.18}$$

Here, θ^n is the contact angle in a time step, θ^{n-1} is the contact angle from previous time step, θ_A is the advancing contact angle, θ_R is the receding contact angle, $\mathbf{u}_{wall}$ is the velocity of fluid near wall surface and $\mathbf{u}_\theta$ is the contact angle velocity scale. In some cases, the interactions between the phases establish a momentum transfer which is very larger than the surface tension acting between them. Such physical processes include transport of solid sedimentations in a river flow, movement of aerosols in air and movement of solid particles during crude oil extraction. For such cases, the momentum transfer between the phases is modelled from force balance of the dispersed phase.

Particle equation of motion

Transfer of momentum between the particle and fluid exists at their interacting region. This momentum transfer results in the dynamic movement of flow. Different forces contribute in momentum transfer between the phases. Stokes [127] developed an initial idea of an unsteady motion of dispersed phase in fluid. Later, Basset [128], Bousinessq [129] and Oseen [130] developed a general equation to describe the nature of forces on a particle submersed in fluid continuum. This proposed Lagrangian equation included the unsteady motion of particle in viscous fluid and was later called BBO equation. Tchen [131] altered BBO equation for the movement of solid particle in a turbulent fluid, by employing a relative velocity between continuous and dispersed phase. This formulation was later corrected by Corrsin and Lumley [132] to calculate the movement of solid particle in a turbulent phase without relative velocity. Maxey and Riley [133] modified the particle equation of motion proposed by Tchen [131]. Neglecting Faxen terms [134], a simple particle equation of motion for a small spherical particle in fluid is

$$m_s \frac{du_s}{dt} = \frac{18\mu_f}{\rho_s d_s^2} m_s(u_f - u_s) - V_s \nabla p + 0.5 m_f \left(\frac{Du_f}{Dt} - \frac{du_s}{dt} \right) + 9\sqrt{\frac{\rho_f \mu_f}{\pi}} \int_0^t \frac{\frac{Du_f}{D\tau} - \frac{du_s}{d\tau}}{(t-\tau)^{0.5}} d\tau + m_s \cdot \mathbf{g}. \tag{4.19}$$

Here, m_s is the particle mass, u_s is the corresponding velocity of solid, μ_f is the dynamic viscosity of continuous phase, ρ_s is the density of dispersed solid phase, d_s is the diameter of solid particle, u_f is the velocity of continuous phase around solid particle, V_s is the volume of solid particle, ∇p is the pressure gradient around a particle in an undisturbed flow with p being the undisturbed pressure field, m_f is the mass of continuous fluid phase, m_s is the mass of dispersed solid particles and τ is the particle response time. Maxey and Riley [133] characterized various forces acting on the particle as different terms in the proposed particle equation of motion. In Eq. 4.19, the first term on right hand side represents drag force on the solid particle. The second term represents pressure gradient in the continuous phase around dispersed phase. The third term describes added mass force and Basset force is implemented in the fourth term on right hand side. The resultant gravitational force acting on particle is implemented as fifth term on right hand side. The term on left hand side represents time rate change of momentum for a solid particle. The pressure gradient term is estimated using Navier-Stokes equation assuming a single-phase fluid as [132]

$$\nabla p = \rho \frac{D\mathbf{u}}{Dt} - \mu \nabla^2 \mathbf{u}. \tag{4.20}$$

This particle equation of motion was developed for particles suspended in flows with low Reynolds numbers. Therefore, Kim et al. [135] developed a new equation of motion for particles suspended in flows with higher Reynolds numbers. Depending on the nature of flow and particle, different forces dominate the particle motion. Brandon and Aggarwal [136] investigated the deposition of particle on a square cylinder inside a channel flow. They have concluded that for density ratios above 20, forces other than viscous drag become unimportant. For particle density ratios below 10 the Basset term seems to be also important. These investigations were performed for flows with Reynolds number between 200 and 2000. Similarly, to calculate the sedimentation velocity a quasi-steady state approach for a particle motion can be assumed [137]. Asako et al. [65] developed a new derivation to obtain similar equilibrium equation like the modified BBO equation. The force equilibrium on solid particle in a rectangular cavity was drawn from the standard Navier-Stokes equation. Here, equation of motion of solid-liquid mixture is governed by general momentum equation. Therefore, the solid settling surrounded by fluid in a rectangular cavity is derived by stipulating incompressible Navier-Stokes equation. The simplified form of this force equilibrium equation can be written as [138]

$$\rho_s \frac{d\mathbf{u}_s}{dt}\, dV_s = \int_S \nabla(\mu_f \nabla \mathbf{u}_f) dS - \int_S \nabla p dS - \int_{V_s} (\rho_s - \rho_f) \mathbf{g}\, dV_s. \tag{4.21}$$

Here, S is the solid surface area, ρ_s is the solid density and ρ_f is the fluid density. By rearranging Eq. 4.21, the resultant force can be written as

$$\mathbf{F} = -\rho_s \frac{d\mathbf{u}_s}{dt}\, dV_s + \int_S \nabla(\mu_f \nabla \mathbf{u}_f) dS - \int_S \nabla p dS - \int_{V_s} (\rho_s - \rho_f) \mathbf{g}\, dV_s. \tag{4.22}$$

Here, $\mathbf{F}$ is the resultant force. The quasi-steady form of settling velocity is calculated using Newton-Raphson iterative approach [65]. Thereby, solid settling velocity $\mathbf{u}_s$ is obtained by

$$\mathbf{u}_s^{n+1} = \mathbf{u}_s^n - \left[\frac{\mathbf{u}_s^{n+1} - \mathbf{u}_s^n}{\mathbf{F}_s^{n+1} - \mathbf{F}_s^n} \right] \times \mathbf{F}^n \tag{4.23}$$

The solid settling velocity is computed by employing Newton-Raphson in every time step. Eq. 4.21 is a simpler representative form of the particle equation of motion as in Eq. 4.19. The drag force in Eq. 4.19 is accounted differently in Eq. 4.21.

4.3 Coupled overlapping grid approaches

Solving day to day multi-physics problems demand combining different physical problems. Physical phenomena like heat transfer including fluid flow, fluid flow with chemical reactions, fluid-structure interaction are few examples of such multi-physics problems. Multi-physics is evolving as an independent computational discipline with growing demand for modelling and simulation in scientific processes. For such problem definitions, different set of governing equations need to be solved for each physical phenomenon. However, these governing equations may or may not interact with each other depending on the nature of the problem definition. Multi-physics problems with interacting governing equations are complex as they demand taking coupling of physical phenomena into account. The complexity of problem definition may increase further when various components made of different materials are interacting with each other. Multi-physics problems sometimes also involve different component interactions. For such computational problems, separate grids are employed for each individual component [139]. Different set of partial differential equation are solved for each grid including interaction between the components. Physical interaction between the components are modelled by overlapping computational grids on another. The overlapping region enables information transfer between independent physical models of the components. An initial idea on dealing with such interacting grid problems was introduced by Starius [140]. In this work, a composite grid was introduced to deal with hyperbolic differential equations with different physical phenomena. Later, a multi-component approach was developed by Atta and Vadyak [139, 141], where component interactions are modelled using grid overlapping schemes. This method was developed to analyse the external fluid flow around a solid aeroplane wing. This numerical method has shown very promising results when compared with pre-existing models. Berger and Oliger [142] have studied the grid refinement at interfaces for grid overlapping schemes with hyperbolic partial differential equations. Henshaw [143] has also dealt with composite mesh approaches with an overlapping mesh interface. In this approach a composite grid G is formed with different separate grids G_i, where i denotes the number of grids involved in a composite mesh and can take a value from 1 to n. This can be mathematically represented as

$$G = G_1 \cup G_2 \cup G_3 \ ......, \tag{4.24}$$

where each mesh component G_i, $i \in$ (1, n) consists of separate grid [143]

$$G_i = x_i(i, j). \tag{4.25}$$

Here, $x_i(i, j)$ are the grid points of computational grid G_i. Mastin [144] reported a stepwise algorithm to solve multi-component overlapping grid problems for one-dimensional transport equation. Browing et al. [145] employed overlapping composite grid approach to compute the rotation of earth with two tangent planes. Chesshire and Henshaw [146] solved different sets of partial differential equations on various overlapping composite grid domains. In the same year, Fuchs [147] employed overlapping grid approach to compute the interaction of solid bodies immersed in a flow. Later, an important contribution from Tu and Fuchs [148] enabled extension of overlapping grid approaches to the field of thermodynamics. In this work, a finite volume method was employed to model fluid flow in a composite overlapping grid of an internal combustion (IC) engine, as seen in Fig. 4.8. Here, the heat transfer and thermal behaviour in an IC engine was neglected.

Adaptive overlapping [149] and moving body overset [150] methods were also some important contributions to the field of overlapping grid approaches. Accurate method for solving the unsteady incompressible Navier-Stokes equations till the order of four using overlapping grids was introduced by Henshaw [151].

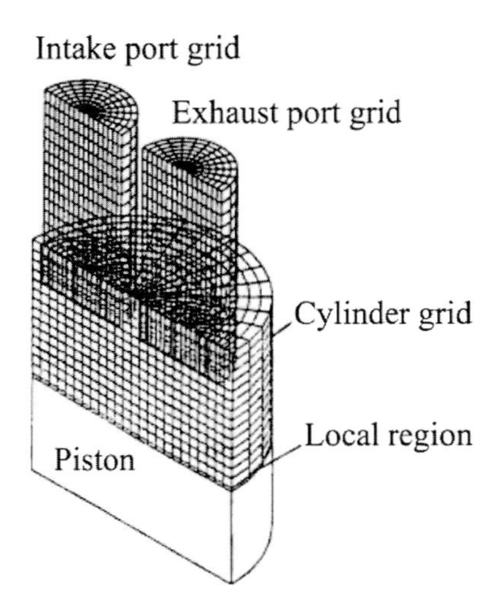

Figure 4.8: Composite overlapping grid of an IC engine with individual grids for each component [148].

Kao and Liou [152] extended the application of coupled overlapped grids for solving compressible flow involving heat transfer. In this work a conjugate heat transfer between fluid and solid interface was modelled, where convective-conductive equation was solved for the fluid domain and in the solid domain simple heat conduction was solved. Later, mapping the data from one grid to another was introduced [153]. Azimi et al. [154] developed a multi-region approach to compute an inverse heat conduction problem by employing overlapping grid approaches. Here, the computational domain is segregated into blocks where the data transfer takes place. A numerical model involving fluid-structure interaction for coupled multi-physics problems was developed later [155]. Interactions at the fluid-solid interface are taken into account by coupling individual set of governing equations for fluid and solid. Here, an unsteady mixed Robin boundary condition was implemented on the fluid-solid interface. This numerical model developed for a flat plate heat exchanger problem was validated with analytical results. Mangani et al. [156] implemented multi-block coupling approach to solve unsteady conjugate heat transfer problem. The inter-facial heat transfer is modelled by inducing time-dependent heat flux between fluid and solid domains. A similar numerical model was implemented in OpenFOAM with an implicit coupling of multi-region blocks in a composite grid [157]. Here, each computational grid for solid or fluid is denoted as a block, where different set of governing equations are solved for each block or region. For a steady state conjugate heat transfer problem, the energy equation for fluid reads

$$\nabla \cdot \left(\mathbf{u}T_f\right) - \nabla \cdot \kappa_f \left(\nabla T_f\right) = 0 \tag{4.26}$$

and the energy equation for the solid domain reads

$$-\nabla \cdot \kappa_s \left(\nabla T_s\right) = 0. \tag{4.27}$$

The heat transfer between fluid and solid domain is modelled using the continuity of temperature. Therefore, at the fluid-solid interface

$$
\begin{aligned}
T_s &= T_f, \\
\kappa_s \frac{\partial T_s}{\partial n} &= -\kappa_f \frac{\partial T_f}{\partial n}.
\end{aligned}
\tag{4.28}
$$

Here, equations and fields with subscript f are solved on the fluid domain. Similarly, equations and fields with subscript s are solved on the solid domain. The interface is an overlapping region between solid and fluid domains.

Volume averaged homogenization

Composite materials are found in different instances from simple everyday ones to complex applications. Composite materials are formed from the combination of two or more materials. The behaviour of composite materials depends on the nature and proportion of each inclusion. Composite materials are commonly heterogeneous due to their constituting materials. Some industrial and natural applications benefit from the heterogeneous structure of composite materials. Investigating such heterogeneous systems using experimental and numerical methods is empirical. Heterogeneity can also be referred to systems with different non-identical components interacting with each other. In industrial applications, almost all systems and machines are heterogeneous in nature as they constitute non-identical components. Numerical modelling of the such heterogeneous systems is arduous, because it is difficult to mathematically represent the complex heterogeneity. Therefore, some simplified approaches exist to deal with complex heterogeneous systems. Volume averaging approach is one of the most common approaches used to represent a heterogeneous system. Here, volume averaging follows representation of a heterogeneous system in a small volume, which describes the heterogeneity in a system. This small volume is called the representative elementary volume (REV). The term REV is termed by Geertsma et al. [158], to estimate the fluid pressure in porous rocks. With REV, the total heterogeneous volume of the system can be treated as a continuum. Hill [159] states that the representative volume can be employed when

- the system is structurally entirely typical of the whole mixture on average and
- contains a sufficient number of inclusions for the apparent overall moduli to be effectively independent of surface values, so long as these values are macroscopically uniform.

Thereby, REV can be defined as a sample used to represent the regular heterogeneity of a larger system [160]. Initially, the volume averaging approach was employed to represent heterogeneous composite material models to compute the mechanical behaviour [161, 162]. While deriving the Darcy law for anisotropic porous media, Neuman [163] reported that Hall [164] and Hubbert [165] have also employed a representative volume approach to average the Navier-Stokes equation for the flow in porous medium. Following Bear [166] and Neuman [163] the intrinsic averaging procedure of a scalar quantity ϕ for anisotropic porous medium reads

$$
\langle \phi \rangle = \frac{1}{V} \int\limits_V \phi dV.
\tag{4.29}
$$

Here, V is the REV of porous medium. From Eq. 4.29, the total heterogeneous system is homogenized with REV as a reference. Therefore, this method is also called the homogenization approach [167]. With the help of volume averaging approach, multiphase composite systems are modelled as homogeneous volume averaged distribution [168]. Besides modelling multiphase composite systems, the volume averaging approach is also very useful to model porous flows [169]. Volume averaging method enables modelling of flow in an irregular porous media, where local averaged transport equation over an REV is computed for a homogenized porous medium. Liu [170] developed a single phase numerical model to represent the mass and momentum transfer in a fluid-solid porous flow. Later, Whitaker [171] introduced a numerical approach to model mass, momentum and heat transport in porous medium. This numerical model was later extended to model the immiscible flow of two fluids in a rigid porous medium [172].

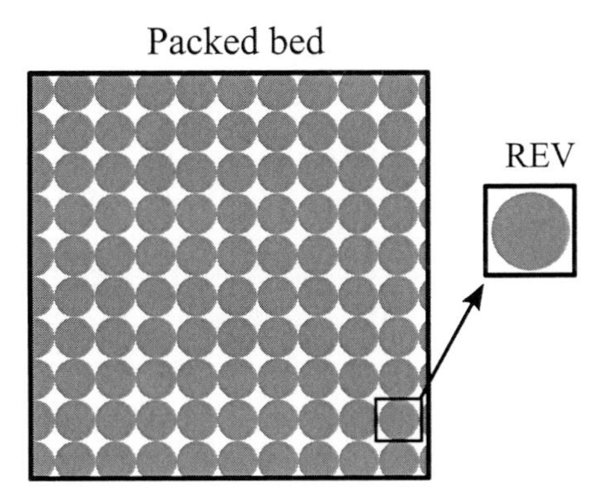

Figure 4.9: Packed bed with spherical bed particles, where a single bed particle in unit cell is the REV.

Numerical model employing the averaging approach was later used to compute heat transfer in packed bed catalytic reactor [173]. The two phases in packed bed catalytic reactor, namely fluid and solid phase were considered to have same temperatures resulting in local thermal equilibrium (LTE). In this work, two different energy equations were solved for both fluid and solid phase. For heterogeneous systems with different length scales, homogenization must also take the varying length scales into account. Therefore, Plumb and Whittaker [174] have proposed two different methods for local averaging of the heterogeneous porous media with different length scales. Amiri and Vafai [175] employed a numerical method to model fluid flow and heat transfer for LTE in a fluid-solid packed bed. In this work, the packed bed is isotropic and consists of solid spheres with uniform shape. Following the volume averaging technique, for a steady-state incompressible flow in packed bed as seen in Fig. 4.9 with constant density, the governing equations read as follows: [175]

Conservation of mass:

$$\nabla \cdot \langle \mathbf{u} \rangle = 0, \tag{4.30}$$

Conservation of momentum:

$$\frac{\rho_f}{\varepsilon}\left\langle(\mathbf{u}\cdot\nabla)\mathbf{u}\right\rangle = \frac{\mu_f}{K}\left\langle\mathbf{u}\right\rangle - \frac{\rho_f F}{\sqrt{K}}\left[\left\langle\mathbf{u}\right\rangle\cdot\left\langle\mathbf{u}\right\rangle\right] + \frac{\mu}{\varepsilon}\nabla^2\left\langle\mathbf{u}\right\rangle - \nabla\left\langle p_f\right\rangle, \tag{4.31}$$

Conservation of energy in the fluid phase:

$$\varepsilon\left\langle\rho_f\right\rangle c_f\frac{\partial\left\langle T_f\right\rangle}{\partial t} + \left\langle\rho_f\right\rangle c_f\left\langle\mathbf{u}\right\rangle\cdot\nabla\left\langle T_f\right\rangle = \nabla\cdot\left[\kappa_{f,\,eff}\cdot\nabla\left\langle T_f\right\rangle\right] + h_{sf}a_{sf}\left[\left\langle T_s\right\rangle - \left\langle T_f\right\rangle\right], \tag{4.32}$$

Conservation of energy in the solid phase:

$$(1-\varepsilon)\left\langle\rho_s\right\rangle c_s\frac{\partial\left\langle T_s\right\rangle}{\partial t} = \nabla\cdot\left[\kappa_{s,\,eff}\cdot\nabla\left\langle T_s\right\rangle\right] - h_{sf}a_{sf}\left[\left\langle T_s\right\rangle - \left\langle T_f\right\rangle\right]. \tag{4.33}$$

Here, $\langle\,\rangle$ denotes volume averaged value of a physical quantity. In the above equations, ρ_f is the density of fluid phase, ε is the porosity or void fraction in a packed bed, μ_f is the dynamic viscosity of fluid phase, p_f is pressure in the fluid phase, K is the permeability of packed bed and F is the Forchheimer constant of packed bed. The permeability of packed bed is dependent on particle distribution and packing density. Thereby, the permeability can be implemented in the numerical model with constants K and F, which read

$$\begin{aligned} K &= \frac{\varepsilon^3 d_p^3}{150(1-\varepsilon)^3}, \\ F &= \frac{1.75}{\sqrt{150\varepsilon^3}}. \end{aligned} \tag{4.34}$$

Here, K depends on the linear pressure difference in axial direction of packed bed. Following Ergun [176], F represents non-linear pressure drop in the packed bed. In Eq. 4.32, Eq. 4.33, c_f is the specific heat capacity of fluid phase, T_f is the temperature in fluid phase and $\kappa_{f,\,eff}$ is the effective thermal conductivity in fluid phase. Accordingly, ρ_s is the density of bed particles in a packed bed, c_s is the specific heat capacity of solid phase, T_s is the temperature in solid phase and $\kappa_{s,\,eff}$ is the effective thermal conductivity in solid phase. The porosity ε in Eq. 4.30, Eq. 4.31, Eq. 4.32 and Eq. 4.33 is defined as the ratio between volume of void spaces in packed bed and the total volume of packed bed.

$$\varepsilon = \frac{V_v}{V_B}. \tag{4.35}$$

Here, V_v is the volume of void spaces and V_B is the volume of packed bed. Amiri and Vafai [175] employed the correlation from Wakao et al. [177, 178] to calculate heat transfer coefficient and mean surface area between fluid and solid particles. Following Wakao et al. [178], for a packed bed with spherical solid particles the heat transfer coefficient between solid and fluid phases h_{sf}

reads

$$h_{sf} = \kappa_f \frac{\left[2 + 1.1\mathrm{Pr}^{\frac{1}{3}} \left(\frac{\rho_f u d_p}{\mu_f}\right)^{0.6}\right]}{d_p}. \tag{4.36}$$

Specific surface area a_{sf} of the bed is formulated as

$$a_{sf} = \frac{6(1-\varepsilon)}{d_p}. \tag{4.37}$$

Here, d_p is the diameter of spherical particle and u is the magnitude of velocity around a spherical particle in bed. This numerical model for steady state flows is further extended to compute unsteady flow and heat transfer in packed beds [179]. The extended numerical model is capable to compute unsteady processes like the charging and discharging of thermal storage [180]. Because, till then either a number of transfer units (NTU) method [181, 182] or simplified porous media approach [183] have been employed to compute charging and discharging of thermal energy storages. Due to the thermal non-equilibrium between fluid and packed bed in a thermal storage, it is not possible to employ a LTE model [184]. Therefore, Sözen and Vafai [185] introduced a numerical model taking the non-thermal equilibrium in a thermal storage unit into account. This numerical model was also called the local thermal non-equilibrium (LTNE) model [186]. Nevertheless, Ismail and Henrıquez [187] still employed a simple one-dimensional model to investigate the behaviour of packed bed latent heat storage system with spherical capsules. Later, Arkar and Medved [188] also used a similar simplified method to compute the thermal response of a LHTES unit. A detailed NTU model was presented by Regin et al. [189]. Here, the thermal response of a single transfer unit was employed to model the total thermal behaviour of a storage unit. Later, Xia et al. [190] employed a conjugate heat transfer model to compute the charging and discharging of a storage unit. They also took into account the natural convection inside the spherical capsules with the help of a steady effective thermal conductivity. Due to the random solid packing, the model does not take the dimensional effects into account. In recent years, some other simplified models [191, 192] were developed to study the transient thermal behaviour of thermal energy storage unit.

All these state of the art methods study physical processes at different length scales. A systematic approach to study the behaviour of macro-encapsulated LHTES unit at different length scales was not found. Alongside studying different processes, the interpolation of parameters from one length scale to another is also important. Such a work was not executed, which draws motivation for the present work.

5 Numerical Modelling

Numerical modelling and simulation contributes to a major part of the present work. This chapter of the work gives an overview of the governing equations and algorithms involved in the numerical modelling. Numerical modelling of physical processes from 20-250 mm capsule scale to 1000-30,000 mm storage unit scale is presented here. Depending on the necessity, both simplified and detailed approaches are employed to represent these processes at different scales.

5.1 Modelling solid-liquid phase change in a storage capsule

In latent heat thermal energy storage units, the solid PCM stores thermal energy in latent form by melting and delivers it back through solidification. Existing empirical methods to model the solid-liquid phase change are discussed in the chapter 3 of this work. In this section of work, a very detailed modelling of solid-liquid phase change in an encapsulated PCM is discussed. During encapsulation, a small air cavity is allocated in the PCM capsule. The air cavity prevents the rupture of PCM capsule due to PCM volumetric expansion during melting. Therefore, the numerical representation of the PCM capsule is represented as a multiphase problem with PCM and air phases. In this work, instead of the standard enthalpy-porosity method, optimum approach is employed to model the solid-liquid phase change. During melting of solid PCM, the solid PCM settling due to density difference is modelled using Newton-Raphson method. Furthermore, the surface interactions between the liquid PCM and gaseous air phase are also taken into account.

5.1.1 Multiphase modelling of a storage capsule

In the multiphase model, the physical problem is mathematically represented using the VOF method. Correspondingly, the total material flux is segregated into PCM and air flux in the numerical model. The flux segregation follows a volume averaging technique, which is represented by the volume fraction of each phase. Accordingly, the volume fraction of any phase i can be defined as

$$\alpha_i = \frac{V_i}{V}. \tag{5.1}$$

A total flux ϕ in the computational domain can be segregated into different fluxes in multiple phases n as

$$\phi = \sum_{i=1}^{n} \alpha_i \phi_i. \tag{5.2}$$

where index $[\ ]_i$ represents the property of a phase i. The conservation of volume fraction α_i between the phases is achieved by

$$\sum_{i=1}^{n} \alpha_i = \alpha_1 + \alpha_2 + \alpha_3 + ... n = 1. \tag{5.3}$$

In the numerical model, material properties and physical quantities can be segregated with their volume fraction in the computational domain. In a PCM capsule with PCM and air phase, the volume fraction of PCM α_1 is defined as

$$\alpha_1 = \frac{V_{PCM}}{V_{Cap}}, \tag{5.4}$$

in the numerical model. Where V_{PCM} is the volume of PCM in the capsule and V_{Cap} the capsule volume. Likewise, the volume fraction of air in a PCM capsule is defined as

$$\alpha_2 = \frac{V_{Air}}{V_{Cap}} = 1 - \alpha_1. \tag{5.5}$$

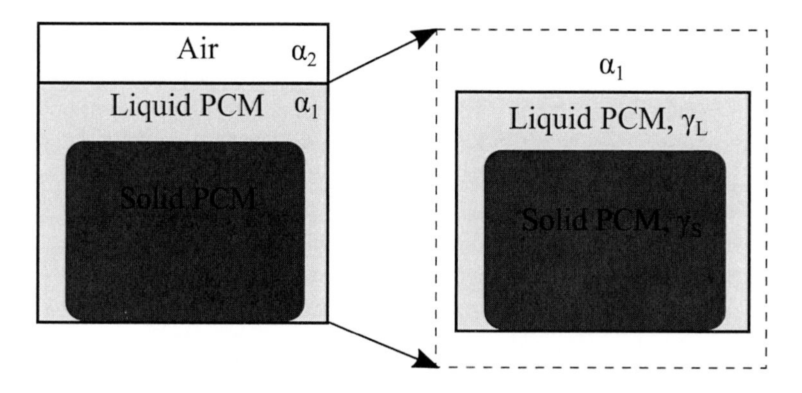

Figure 5.1: Pictorial representation of each phase in a multiphase PCM capsule.

The solid-liquid phase change in the PCM capsule results in two sub-phases of PCM phase i.e. the solid PCM phase and the liquid PCM phase. The volume fraction of solid PCM phase γ_S inside the PCM-phase is defined as

$$\gamma_S = \frac{V_{PCM,\,S}}{V_{PCM}} \tag{5.6}$$

and the liquid PCM phase γ_L of the PCM-phase is defined as

$$\gamma_L = \frac{V_{PCM,\,L}}{V_{PCM}}. \tag{5.7}$$

Here, $V_{PCM,\,S}$ is the volume of solid PCM and $V_{PCM,\,L}$ is the volume of liquid PCM.

Correspondingly, the total flux ϕ in the multiphase PCM capsule can be written as

$$\phi = \alpha_1 \phi_1 + \alpha_2 \phi_2, \tag{5.8}$$

which can also be sub-segregated as

$$\phi = \alpha_1 (\gamma_L \phi_L + \gamma_S \phi_S) + \alpha_2 \phi_2. \tag{5.9}$$

Here, the index $[\ ]_L$ represents the liquid PCM phase, $[\ ]_S$ is the solid PCM phase, $[\ ]_1$ denotes the PCM phase and $[\ ]_2$ denotes the air phase in the PCM capsule.

Multiphase physical processes involving fluid flow and heat transfer is the topic of interest here. For multiphase flows, the standard governing equations like Eq. 3.8, Eq. 3.9 and Eq. 3.10 used to represent the heat transfer and fluid flow are also segregated between the phases. Following Rusche [110], the velocity can be segregated between the PCM and air phase as

$$\mathbf{u} = \alpha_1 \mathbf{u}_1 + \alpha_2 \mathbf{u}_2. \tag{5.10}$$

From the Eq. 3.8, the continuity equation for the mass conservation of multiphase flow in a control volume can defined as

$$\frac{D\rho}{Dt} = \frac{D}{Dt}(\alpha_1 \rho_1 + \alpha_2 \rho_2) = 0. \tag{5.11}$$

For compressible flows, the density ρ is not constant and by rearranging the continuity is represented as

$$\frac{\partial \alpha_1 \rho_1}{\partial t} + \nabla \cdot (\alpha_1 \rho_1 \mathbf{u}) + \alpha_1 \left[\frac{\partial \rho_1}{\partial t} + \nabla \cdot (\rho_1 \mathbf{u}) \right] = 0 \tag{5.12}$$

for the PCM phase and

$$\frac{\partial \alpha_2 \rho_2}{\partial t} + \nabla \cdot (\alpha_2 \rho_2 \mathbf{u}) + \alpha_2 \cdot \left[\frac{\partial \rho_2}{\partial t} + \nabla \cdot (\rho_2 \mathbf{u}) \right] = 0 \tag{5.13}$$

for the air phase. The multiphase continuity equation Eq. 5.11 is segregated into two different equations as shown in Eq. 5.12 and Eq. 5.13. In a PCM capsule, the PCM phase is treated as a incompressible fluid and the air phase as compressible fluid. For compressible flow, the density change is implemented in OpenFOAM [193] using the dynamic pressure change. Thereby, the continuity equation for air phase in the present numerical model is written as

$$\frac{\partial \alpha_2 \rho_2}{\partial t} + \nabla \cdot (\alpha_2 \rho_2 \mathbf{u}) + \alpha_2 \left[\frac{\partial \rho_2}{\partial t} + \nabla \cdot (\rho_2 \mathbf{u}) \right] + \psi_2 \frac{\partial p_d}{\partial t} = 0, \tag{5.14}$$

where ψ_2 is the compressibility of the air phase. As the PCM is incompressible, the additional pressure influenced compressibility is excluded in the continuity equation of PCM phase. As described in the Eq. 4.12, the corrected phase continuity equation can be adapted for the two-phase compressible flows and can be written as

$$\frac{\partial \alpha_1}{\partial t} + \nabla \cdot (\alpha_1 \mathbf{u}) + \nabla \cdot (\alpha_1 \alpha_2 \mathbf{u}_r) = -\frac{\alpha_1}{\rho_1} \frac{D\rho_1}{Dt} \tag{5.15}$$

and

$$\frac{\partial \alpha_2}{\partial t} + \nabla \cdot (\alpha_2 \mathbf{u}) + \nabla \cdot (\alpha_1 \alpha_2 \mathbf{u}_r) = -\frac{\alpha_2}{\rho_2} \left[\psi_2 \frac{\partial p_d}{\partial t} \right]. \tag{5.16}$$

Here, $\mathbf{u}_r$ is the relative velocity between the phases equalling to $\mathbf{u}_1 - \mathbf{u}_2$. According to Rusche [110], the corrected phase continuity equation with an additional term ensures the boundedness of volume fraction α_1 in the computational domain.
Conservation of momentum in the multiphase system is represented by a single mixture equation as in Eq. 3.9. Accordingly, momentum equation for the multiphase PCM-Air system is represented in the numerical model as

$$\frac{\partial \rho \mathbf{u}}{\partial t} + \nabla \cdot (\rho \mathbf{u}\mathbf{u}) = -\nabla p + \mu \Delta \mathbf{u} + \rho \cdot \mathbf{g} + \sigma \kappa(\gamma_L) \nabla \alpha_1. \tag{5.17}$$

According to Pope [194], the total pressure p is segregated as

$$p = p_d + \rho \mathbf{g} \cdot \mathbf{x} \tag{5.18}$$

and its gradient is defined as

$$\nabla p = \nabla p_d + \mathbf{g} \cdot \mathbf{x} \nabla \rho + \rho \mathbf{g}. \tag{5.19}$$

By substituting Eq. 5.19 in Eq. 5.17 the momentum equation becomes

$$\frac{\partial \rho \mathbf{u}}{\partial t} + \nabla \cdot (\rho \mathbf{u}\mathbf{u}) = -\nabla p_d - \mathbf{g} \cdot \mathbf{x} \nabla \rho + \mu \Delta \mathbf{u} + \sigma \kappa_{sf}(\gamma_L) \nabla \alpha_1. \tag{5.20}$$

Here, p_d is the modified pressure or the dynamic pressure, σ is the coefficient of surface tension, $\kappa_{sf}(\gamma_L)$ is the curvature between the liquid PCM and the air phase and $\mathbf{x}$ is the coordinate vector. The curvature between the liquid PCM and air phase as seen in Fig. 5.2 depends on the advection of liquid PCM phase, which is defined as

$$\kappa_{sf}(\gamma_L) = -\left[\nabla \cdot (\mathbf{n} \cdot \mathbf{S}_f) \right] \gamma_L \tag{5.21}$$

and the normal vector of interface $\mathbf{n}$ is defined as

$$\mathbf{n} = \frac{\nabla \alpha_1}{|\nabla \alpha_1|}. \tag{5.22}$$

The interface curvature between liquid PCM and air forms a contact angle with the capsule wall. To account for the change in the curvature of the interface between liquid PCM and air, a contact angle is implemented as a boundary condition.

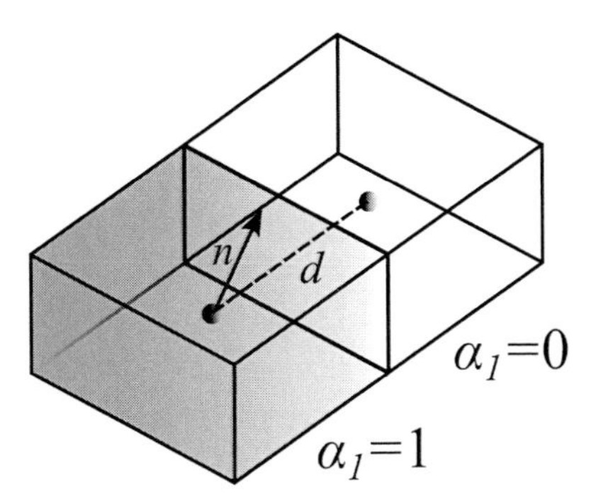

Figure 5.2: Pictorial depiction of PCM volume fraction progression with normal vector n.

The product of interface normal vector $\mathbf{n}$ and face normal vector at the wall $\mathbf{n}_w$ is balanced in every time step as

$$\cos\theta = \mathbf{n}_w \cdot \mathbf{n}. \tag{5.23}$$

The contact angle θ coinciding with the capsule wall, liquid PCM and air is defined as

$$\cos\theta = \frac{\sigma_{WG} - \sigma_{WL}}{\sigma_{LG}}. \tag{5.24}$$

Here, σ_{WG} is the surface tension coefficient between the capsule wall and the gaseous air phase, σ_{WL} is the surface tension coefficient between the capsule wall and liquid PCM phase and σ_{LG} is the surface tension coefficient between the gaseous air phase and the liquid PCM phase. The contact angle θ generally varies depending on the material type and their wettability. Different contact angles exist depending on the nature of wettability, where a contact angle $\theta = 0\ ^\circ$ defines a perfect wetting state and $\theta = 180\ ^\circ$ defines a perfect non-wetting condition.

In the numerical model, material properties of multiphase system are also segregated using volume fraction between PCM-Air phases as in Eq. 5.9. The dynamic viscosity μ of the multiphase mixture is defined as

$$\mu = \alpha_1\mu_1 + \alpha_2\mu_2, \tag{5.25}$$

where μ_1 is the dynamic viscosity of PCM phase and μ_2 is the dynamic viscosity of air. In the PCM phase, a large dynamic viscosity is implemented in the solid PCM phase to represent its non-fluidic nature. Here, the numerical model employs the VVM to represent the suppression of velocity in solid PCM phase. Therefore, the solid PCM in the numerical model is considered as solidius liquid with a large viscosity value. The dynamic viscosity in the PCM phase μ_1 follows an exponential relationship with temperature defined as

$$\mu_1 = \mu_L \cdot \left[1 - \beta_\mu \Delta T\right] \cdot e^{A\gamma_S}. \tag{5.26}$$

Here, A is the exponential coefficient, μ_L is the dynamic viscosity of the liquid PCM, β_μ is the viscosity expansion coefficient and ΔT is the temperature change. The Bousinessq approximation is employed to account density change in the numerical model. Therefore, in the Eq. 5.20 the PCM-Air mixture density ρ is implemented as

$$\rho = \alpha_1 \left[\gamma_S \rho_S + \gamma_L \rho_L \cdot (1 - \beta \Delta T)\right] + \alpha_2 \rho_2, \tag{5.27}$$

where, ρ_L is the density of liquid PCM, ρ_S is the density of solid PCM, β is the coefficient of thermal expansion, ΔT is the temperature change and ρ_2 is the density of air. In the present numerical model, mass and momentum conservation are strongly dependent on the phase change process. The numerical model takes the non-linear enthalpy-temperature into account to describe solid-liquid phase change in the PCM phase. As in Eq. 5.8, the enthalpy of multiphase mixture h in the PCM capsule can be defined as

$$h = \alpha_1 h_1 + \alpha_2 h_2, \tag{5.28}$$

where h_1 contributes to the enthalpy of PCM, which is non-linearly related to the temperature and h_2 is the enthalpy in the air phase. The non-linear enthalpy-temperature relationship of PCM phase as depicted in Fig. 4.2 is characterized by its non-linear slope $\frac{dh}{dT}$. Thereby, from Eq. 5.28 the change in enthalpy can be written as

$$dh = \alpha_1 \left[\frac{dh}{dT}\right]_1 \cdot dT + \alpha_2 \left[\frac{dh}{dT}\right]_2 \cdot dT. \tag{5.29}$$

The ratio of the enthalpy change within the change in temperature is the apparent heat capacity c^A, which is also the slope of the enthalpy-temperature curve $\frac{dh}{dT}$. Therefore, by replacing the slope in Eq. 5.29 with the apparent heat capacity c^A and for the air phase with heat capacity c_2, we obtain

$$dh = \alpha_1 \left[c^A\right] \cdot dT + \alpha_2 c_2 \cdot dT. \tag{5.30}$$

The non-linear slope of the enthalpy-temperature plot can be defined either as a piecewise linear function or polynomial. In the present work, a Gaussian normal distribution is employed to represent the apparent heat capacity as

$$c^A = \left[\frac{dh}{dT}\right]_1 = \left[\gamma_S c_S + \gamma_L c_L\right] + \frac{L}{\sqrt{2\pi}\sigma_m} \cdot e^{-0.5\left[\frac{T - T_m}{\sigma_m}\right]^2}. \tag{5.31}$$

Here, c_S is the specific heat capacity of solid PCM, c_L is the specific heat capacity in the liquid phase and σ_m is the standard distribution of the Gaussian normal distribution defined as

$$\sigma_m = \frac{T_L - T_S}{4}. \tag{5.32}$$

Here, the optimum approach [51] proposed by Swaminathan and Voller is employed to model the non linear enthalpy-temperature relation.

Optimum approach

Swaminathan and Voller [50, 51] have proposed an iterative method to solve solid-liquid phase change problems with low computational time.

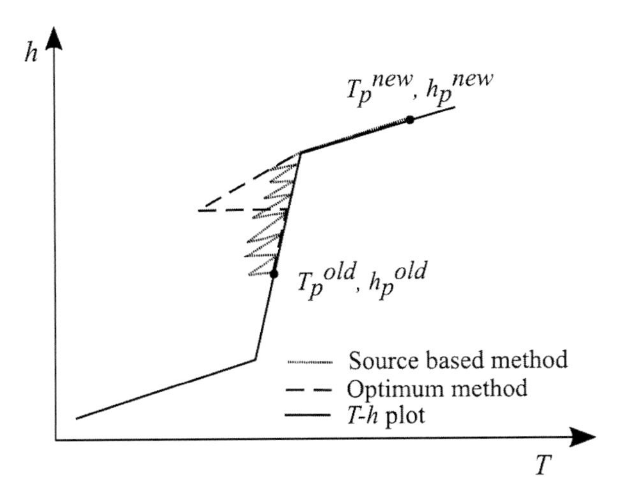

Figure 5.3: Comparison between the optimum approach [50] and the source based method [39] to model the non-linear T-h plot.

In this method, the enthalpy is defined as a function of temperature. The simplicity of the method lies in its instantaneously accessible slope $\frac{dh}{dT}$ at every point of T-h curve. The inverse function of the enthalpy h^{-1} denotes the temperature for corresponding enthalpy h in the T-h plot. To avoid discontinuities in the curve, Swaminathan and Voller [50] proposed selection of a large arbitrary value for $\frac{dh}{dT}$ inside the melting range. For a conductive phase change problem, Voller [56] claims that in most of the time steps the optimum approach requires only a single iteration step to converge.

Therefore, in this work the optimum approach has been used to model the solid-liquid phase change. The slope of enthalpy-temperature curve for the PCM phase is defined as in Eq. 5.31. For a multiphase PCM capsule with PCM-Air a compressible set of governing equations are considered. Ignoring the influence of pressure and viscous dissipation, the simplified form of energy equation for the compressible flow defined as in Eq. 3.10 is

$$\frac{\partial \rho h}{\partial t} + \nabla \cdot (\rho \mathbf{u} h) = \nabla \cdot (\kappa \nabla T). \tag{5.33}$$

By implementing Eq. 5.30 in Eq. 5.33, the energy conservation equation of multiphase PCM-Air mixture in a PCM capsule becomes non-linear. Following Swaminathan and Voller [51], the discretized form of energy conservation equation can represented as

$$a_p T_p^{m+1} + d_p \left[c^A\right]_{h_p^m} \left(T_p^{m+1} - \left[h_p^m\right]^{-1}\right) = \sum_{nb} a_{nb} T_{nb}^{m+1} + d_p \left[h_p^o - h_p^m\right] + b. \tag{5.34}$$

Here, a_p, b and d_p are the discretization coefficients, o is the old time step, nb is the subscript defining neighbouring cells. For an iterative approach, $m+1$ is the present iterative step and m is the previous iterative step. According to Swaminathan and Voller [51], an implicit enthalpy can be calculated using the truncated Taylor series as

$$h_p^{m+1} = h_p^m + \omega \left[c^A\right]_{h_p^m} \cdot \left(T_p^{m+1} - T_p^m\right). \tag{5.35}$$

Here, ω is the under relaxation factor. Following this, the temperature is corrected from the updated enthalpy. For this an inverse function of enthalpy is calculated which is written as

$$\left(h^{-1}\right)_p^{m+1} = \begin{cases} \dfrac{h_p^{m+1}}{c_S}, & h_p^{m+1} < c_S T_S \\ T_m + \sigma_m \cdot \tanh^{-1}\left[\dfrac{h_p^{m+1} - c_S T_S}{c_L T_L + L - c_S T_S + (c_S - c_L)T_m}\right], & c_S T_S \leq h_p^{m+1} \\ & \leq c_L T_L + L + (c_S - c_L)T_m \\ \dfrac{h_p^{m+1} - L - (c_S - c_L)T_m}{c_L}, & h_p^{m+1} > c_L T_L + L + (c_S - c_L)T_m. \end{cases} \tag{5.36}$$

Here, $\left(h^{-1}\right)_p^{m+1}$ is equal to T_p^{m+1} at convergence. The volume fraction of the liquid PCM is calculated from the solved temperature as an integral of the apparent heat capacity, which can be written as

$$\gamma_L = 0.5 + 0.5\text{erf}\left[\frac{T - T_m}{\sqrt{2}\sigma_m}\right] \approx 0.5 + 0.5\tanh\left[\frac{T - T_m}{\sigma_m}\right]. \tag{5.37}$$

Within the optimum approach, Eq. 5.34, Eq. 5.35 and Eq. 5.36 are solved and iterated till the convergence is achieved. The convergence of the energy equation is characterized by a convergence criterion ε_{conv}, which is mathematically represented as in Eq. 5.38.

$$|T_p^{m+1} - T_p^m| < \varepsilon_{conv}. \tag{5.38}$$

Close contact melting

Density difference between the liquid and solid phase of PCM leads to a buoyancy influenced settling of denser solid to the bottom of capsule. In the multiphase numerical model, different densities for the solid and liquid PCM were implemented in the mass and momentum conservation equation as seen in Eq. 5.27, Eq. 5.20 and Eq. 5.15. Despite the density difference, the solid settling demands a large computational time achieve a convergence. Therefore, in the present work a force equilibrium method is employed to influence the solid PCM settling in a macro-capsule.

Explicit approach: Force equilibrium method

Following Asako et al. [65], the force equilibrium is calculated by summing up different forces acting on the solid PCM body. The settling velocity u_S is calculated from the force equilibrium using Newton-Raphson method. Instead of six degrees of freedom (DOF) a single degree of freedom is considered to calculate the force equilibrium [79]. The implementation of settling velocity follows a multiphase mixture approach as in Eq. 5.9. The velocity of the mixture is defined as

$$\mathbf{u} = \alpha_1(\gamma_L \mathbf{u} + \gamma_S \mathbf{u}_S) + \alpha_2 \mathbf{u}. \tag{5.39}$$

The velocity $\mathbf{u}$ in the solid phase from Eq. 5.20 is replaced by the settling velocity $\mathbf{u}_s$ retrieved from force equilibrium. By implementing the settling velocity, a relative velocity $\mathbf{u} - \mathbf{u}_s$ exists between the solid and liquid PCM phases. The force equilibrium on the solid PCM is obtained from the sum of forces acting on it like the body force, pressure gradient, drag and inertial forces. Following Rösler [79], the force equilibrium on the solid PCM in a PCM-capsule can be written as the sum of pressure gradient and body forces as

$$\mathbf{F}_{res} = \mathbf{F}_h + \mathbf{F}_g. \tag{5.40}$$

Here, $\mathbf{F}_{res}$ is the resulting force obtained by the sum of hydrostatic and gravitational forces. The hydrostatic force $\mathbf{F}_h$ acting on the solid PCM surface area holds up the solid PCM. The change in the pressure around the solid PCM's area is defined by Froude–Krylov force equation for unsteady pressure field [195]. The hydrostatic force acting on the solid PCM surface is described as

$$\mathbf{F}_h = \int_s p \cdot \mathbf{n} \, ds. \tag{5.41}$$

Here, s is the surface area of the solid PCM and $\mathbf{n}$ is the normal vector in the direction of acting pressure field. The body force $\mathbf{F}_g$ is defined as the volumetric weight of the solid phase in the PCM capsule. The volumetric body force is unsteady, which depends on the solid PCM volume. The volumetric body force is defined as

$$\mathbf{F}_g = m_S \cdot \mathbf{g} = \rho_S \, V_S \cdot \mathbf{g}. \tag{5.42}$$

Here, m_S is the mass of solid PCM and V_S is the volume of solid PCM. The resistance force influencing the lift of the solid PCM is the sum of pressure gradient and shear stress force, see [65]. In this work, the influence of the shear stress force on the solid PCM is neglected [79]. Thereby, the resulting force $\mathbf{F}_{res}$ can be written as

$$\mathbf{F}_{res} = m_S \cdot \frac{d\mathbf{u}_S}{dt} = \mathbf{F}_h + \mathbf{F}_g. \tag{5.43}$$

The force equilibrium can be represented in the form [65]

$$\mathbf{F} = -m_S \cdot \frac{d\mathbf{u}_S}{dt} + \int_s p \cdot \mathbf{n} \, ds + \rho_S \, V_S \cdot \mathbf{g} = 0. \tag{5.44}$$

The unsteady change of velocity can be excluded for quasi-steady settling velocities, because it is sufficient to compute a quasi-steady velocity of particle in sedimentation processes [137].

The implicit Newton-Raphson method for an implicit steady settling velocity from Eq. 5.45 is

$$\mathbf{F} = \int_{S} p \cdot \mathbf{n}\, ds + \rho_S\, V_S \cdot \mathbf{g} = 0. \tag{5.45}$$

The force equilibrium in Eq. 5.45 can be solved by an iterative Newton-Rapshon method proposed by Asako et al. [65]. The implicit settling is calculated by solving Eq. 5.46

$$\mathbf{u}_S^{m+1} = \mathbf{u}_S^{m} - \left[\frac{\mathbf{u}_S^{m-1} - \mathbf{u}_S^{m}}{\mathbf{F}^{m-1} - \mathbf{F}^{m}}\right] \cdot \mathbf{F}^{m}. \tag{5.46}$$

Here, $m+1$ contributes to the implicit present iterative step, m denotes the past iterative step and $m-1$ is the post previous iterative step. The loop is iterated till the convergence where the difference between the values of present and previous iterative steps fall below convergence. Due to the implicit iterative solution, the method is a slow converging approach. Therefore for a converged pressure and velocity, Rösler [79] proposed an explicit solution of the Newton-Raphson method for Eq. 5.45. Here, the explicit method to compute the settling velocity follows

$$\mathbf{u}_S = \mathbf{u}_S^{o} - \left[\frac{\mathbf{u}_S^{oo} - \mathbf{u}_S^{o}}{\mathbf{F}^{oo} - \mathbf{F}^{o}}\right] \cdot \mathbf{F}, \tag{5.47}$$

where oo stands for the post-previous time step and o stands for the old time step time. The obtained settling velocity is implemented in the Eq. 5.39.

Solution algorithm

The developed numerical model solves the solid-liquid phase change of PCM under the influence of air phase. Additionally, it includes the compression of air phase influenced by expansion of PCM phase in a capsule, liquid PCM wetting on capsule wall and close contact melting in a PCM capsule. In a time step,

- initially Eq. 5.15 is solved from

  $$\frac{\partial \alpha_1}{\partial t} + \nabla \cdot (\alpha_1 \mathbf{u}) + \nabla \cdot (\alpha_1 \alpha_2 \mathbf{u}_r) = 0 \tag{5.48}$$

  using the MULES (multidimensional universal limiter with explicit solution) algorithm.

- From the solution of α_1, α_2 is computed consequently with Eq. 5.5. Thereby, the mixture continuity in the computational domain is calculated using

  $$\frac{\partial \rho}{\partial t} + \nabla \cdot (\rho \mathbf{u}) = 0, \tag{5.49}$$

  where ρ is the mixture density. Later, the energy conservation equation as in Eq. 5.33 is solved using the optimum approach.

- For the optimum approach, Eq. 5.34, Eq. 5.35 and Eq. 5.36 are solved and iterated till the residual of the energy equations falls below the residual as in Eq. 5.38. The volume fraction of liquid PCM is calculated and updated with Eq. 5.37.

- Later, the momentum equation with the pressure correction is obtained by solving the Eq. 5.50

$$\nabla \cdot \mathbf{u}_P = \nabla \cdot \mathbf{u}_P^* - \nabla \cdot \left[\frac{(\nabla p)_P}{\mathbf{A}_P} \right] = -\frac{\alpha_1}{\rho_1} \left[\frac{D\rho_1}{Dt} \right] - \frac{\alpha_2}{\rho_2} \left[\frac{D\rho_2}{Dt} + \psi_2 \frac{\partial p_d}{\partial t} \right]. \tag{5.50}$$

- The material properties are updated from the obtained pressure, velocity, melting fractions and temperature. The density of the compressible air phase is updated as

$$\rho_2 = \rho_2^{\mathrm{o}} + \psi_2 \cdot (\Delta p_d), \tag{5.51}$$

where ρ_2^{o} is the density of the air from the previous time step. Later, the density of the mixture is updated using the Eq. 5.27.

- For the implementation of the settling algorithm, the numerical model solves the force equilibrium and implements the solid velocity in the mixture velocity as in Eq. 5.39.

- After solving of the settling algorithm, the energy balance is verified between the energy input from walls derived from the Fourier law and the change in enthalpy. The difference of the energy is approximated as numerical error by the algorithm, which is written as [79]

$$\Delta Q = Q_{in} - Q_h = \left(Q^{\mathrm{o}} + \dot{Q}\Delta t\right) - \left(\sum_P h_P V_P \right). \tag{5.52}$$

Here, V_P is the volume of the computing cell with node P, Q is the thermal energy and $\dot{Q}$ is the heat flux inside a control volume which can be written as

$$\dot{Q} = \lambda_f s_f \frac{dT}{d\mathbf{n}}, \tag{5.53}$$

where λ_f is the thermal conductivity in the control volume, s_f is the surface area of the control volume and $\mathbf{n}$ is a normal vector in the direction of temperature flow. The enthalpy at the control node h_P for Eq. 5.52 is calculated as

$$h_P = \rho \left[cT + \alpha_1 \gamma_L \left(L + (c_S - c_L)T_m\right)\right]. \tag{5.54}$$

- With the updated material properties, velocity, temperature and pressure in the computational domain, the solution progresses to the next time step.

These steps are iteratively executed till the end time with an adapted time step of the problem definition.

The present multiphase model is not just complex to implement but also computationally expensive. The numerical model considers all relevant physical phenomena in the PCM capsule during the solid-liquid phase change in detail. But due to its computational costs the model is not suitable for parametric study. Therefore, simplified numerical approaches are developed to represent the solid-liquid phase change in a PCM capsule with reduced computational effort.

5.1.2 Simplified numerical modelling of a storage capsule

For simplified numerical models, governing equations in a simplified manner are solved to estimate the solid-liquid phase change in a PCM capsule. In this section of work, the simplified convective model reduces the complex multiphase system into a single PCM phase model by modelling the air phase as a convective boundary condition. Here, the solid-liquid phase change influenced by heat transfer with solid PCM settling is represented.

A simplified conductive model helps to represent the physical phenomena in a single PCM capsule with large scale computations for a thermal storage unit. With this numerical model, effective thermal conductivity is analysed for a defined PCM and capsule geometry. The validated and analysed effective thermal conductivity for a PCM capsule is then implemented in the LTNE model for a large scale thermal storage unit.

Simplified convective model

As discussed, this numerical model represents the air phase as a heat transfer boundary condition at the PCM-Air interface inside the capsule. The governing equations for a multiphase system are reduced to a single PCM phase.

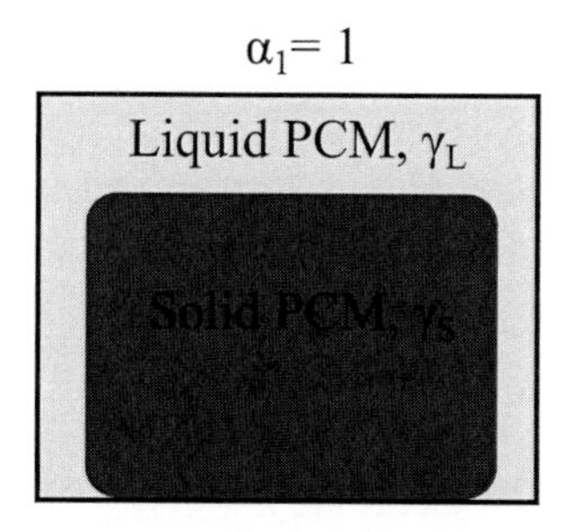

Figure 5.4: Pictorial description of solid-liquid phase change with simplified convective model.

For this purpose, a set of incompressible Navier-Stokes governing equations are employed to model the solid-liquid phase change in PCM. These equations are derived by stipulating the volume fraction for the full PCM phase, by implementing $\alpha_1 = 1$ and $\alpha_2 = 0$ in the multiphase governing equations. Accordingly, the mass conservation of the incompressible flow from Eq. 3.8 is reduced to

$$\nabla \cdot \mathbf{u} = 0 \tag{5.55}$$

for an assumption of constant density. The buoyancy influenced by the density change is modelled using Bousinessq approximation. An appropriate momentum conservation equation in the simplified convective model is written as

$$\rho\left[\frac{\partial \mathbf{u}}{\partial t}+\nabla\cdot(\mathbf{uu})\right]=-\nabla p_d-\mathbf{g}\cdot\mathbf{x}\nabla\rho_i+\nabla\cdot(\mu\nabla\mathbf{u}). \tag{5.56}$$

Where the density change in PCM is written as

$$\rho_i=\gamma_S\rho_S+\gamma_L\rho_L\cdot(1-\beta\Delta T). \tag{5.57}$$

Here, ρ_i is the mixture density of PCM phase with Bousinessq approximation for natural convection in the liquid phase. By employing VVM, velocity in solid phase is suppressed. Using VVM, the dynamic viscosity μ for solid-liquid phase change in PCM can be defined as

$$\mu=e^{A\gamma_S}\mu_L\cdot\left[1-\beta_\mu\Delta T\right]. \tag{5.58}$$

The energy equation in a simplified convective model for a single PCM phase is written as

$$\frac{\partial\rho h}{\partial t}+\nabla\cdot(\rho\mathbf{u}h)=\nabla\cdot(\kappa\nabla T). \tag{5.59}$$

The energy equation as in Eq. 5.59 is solved using the optimum approach [51], where enthalpy h refers to the energy in PCM. Accordingly, the discretized form of the equation follows the same equation as in Eq. 5.34. The reformulation of energy equation with the slope of the enthalpy-temperature curve $\frac{dh}{dT}$ replaced by apparent heat capacity c^A yields

$$c^A\frac{\partial\rho T}{\partial t}+c^A\nabla\cdot(\rho\mathbf{u}T)=\nabla\cdot(\kappa\nabla T). \tag{5.60}$$

The apparent heat capacity c^A is implemented in the equation as defined in Eq. 5.31. The enthalpy in the PCM phase is updated using

$$h_p^{m+1}=h_p^m+\omega\cdot\left[c^A\right]_{h_p^m}\cdot\left(T_p^{m+1}-T_p^m\right). \tag{5.61}$$

Following the enthalpy update, the temperature is corrected for a consistency with enthalpy. The temperature is corrected from the implicit enthalpy h_p^{m+1} as

$$\left(h^{-1}\right)_p^{m+1}=\begin{cases}\dfrac{h_p^{m+1}}{c_S}, & h_p^{m+1}<c_S T_S\\ T_m+\sigma_m\cdot\tanh^{-1}\left[\dfrac{h_p^{m+1}-c_S T_S}{c_L T_L+L-c_S T_S+(c_S-c_L)T_m}\right], & c_S T_S\le h_p^{m+1}\\ & \le c_L T_L+L+(c_S-c_L)T_m\\ \dfrac{h_p^{m+1}-L-(c_S-c_L)T_m}{c_L}, & h_p^{m+1}>c_L T_L+L+(c_S-c_L)T_m.\end{cases} \tag{5.62}$$

Later, the convergence of iterative approach is characterized using Eq. 5.38. The corrected temperature is then used to calculate the vol. fraction of the liquid PCM in the computational domain as shown in Eq. 5.37.

Close contact melting

An explicit or implicit method can be employed to influence the settling of solid PCM in PCM capsule. Following Asako et al. [65], the explicit method is solved using Newton-Raphson method. The settling velocity is solved by force-equilibrium using an explicit approach. The force-equilibrium equation for the incompressible flow assumption of the PCM phase in the capsule is

$$\mathbf{F} = \int_{s} p \cdot \mathbf{n}\, ds + \rho_S\, V_S \cdot \mathbf{g} = 0. \tag{5.63}$$

The explicit settling velocity of the solid PCM obtained by solving the force equilibrium using Newton-Raphson method is

$$\mathbf{u}_S = \mathbf{u}_S^{\mathrm{o}} - \left[\frac{\mathbf{u}_S^{\mathrm{oo}} - \mathbf{u}_S^{\mathrm{o}}}{\mathbf{F}^{\mathrm{oo}} - \mathbf{F}^{\mathrm{o}}} \right] \cdot \mathbf{F}. \tag{5.64}$$

The obtained settling velocity is implemented in the solid PCM phase of the computational domain as

$$\mathbf{u} = \gamma_L \mathbf{u} + \gamma_S \mathbf{u}_S. \tag{5.65}$$

For an implicit approach, iterative pressure-velocity loop is employed to compute the settling velocity. Where, the solid settling velocity $\mathbf{u}_S$ is achieved by implementing the buoyancy force in the momentum conservation equation. Therefore for the present numerical model, the settling velocity is computed by solving

$$\nabla \cdot \mathbf{u}_P^* = \nabla \cdot \left[\frac{(\nabla p)_P}{\mathbf{A}_P} \right]. \tag{5.66}$$

Solution algorithm

The simplified convective model employs governing equations for incompressible flow to account the solid-liquid phase change of PCM including the settling of solid PCM in a capsule. Inside a time step,

- initially momentum equation from Eq. 5.56 is solved to obtain the velocity $\mathbf{u}$ from the initial conditions of velocity, temperature and pressure.

- Later, the energy equation Eq. 5.60 is solved with updated material properties like apparent heat capacity c^A and thermal conductivity κ using the optimum approach. The liquid fraction of the PCM is then calculated using Eq. 5.37. The convergence of the energy equation is governed by Eq. 5.38 following optimum approach.

- The pressure is updated and the velocity is corrected inside a PIMPLE loop.

- Later, the solid settling velocity $\mathbf{u}_S$ is implemented in the Eq. 5.65 by solving Eq. 5.63 and Eq. 5.64.

- The energy balance is verified by the numerical model from Eq. 5.52, while the enthalpy at computing node P is defined as

$$h_P = \rho \left[cT + \gamma_L \left(L + (c_S - c_L) T_m \right) \right]. \tag{5.67}$$

These solution steps are executed inside every time step till the end of computational time. Hereby, the simplified convective model estimates the solid-liquid phase change in PCM phase of a PCM capsule. But, this numerical model is not simple enough to implement it into a large scale thermal storage unit. Therefore, a simplified effective thermal conductivity model is employed to estimate the role of natural convection and solid settling in a PCM capsule.

Simplified effective Conductive model

Here, the numerical model is simplified by representing complex phenomena like natural convection, close contact melting and interfacial heat transfer through an effective thermal conductivity.

Figure 5.5: Depiction of the melting pattern of PCM with the simplified effective conductive model.

For this, a simple conductive energy equation is taken into account, which is written as

$$\frac{\partial \rho h}{\partial t} = \nabla \cdot (\kappa_{eff} \nabla T). \tag{5.68}$$

Here, κ_{eff} is the effective thermal conductivity of PCM which is defined as

$$\kappa_{eff} = \gamma_L \cdot \kappa_{eff,\,L} + \gamma_S \cdot \kappa_S. \tag{5.69}$$

The functional form of energy equation can be represented as

$$c^A \frac{\partial \rho T}{\partial t} = \nabla \cdot (\kappa_{eff} \nabla T) \tag{5.70}$$

where the solid takes an unaltered value of thermal conductivity κ_S due to its independence from natural convection. The liquid phase of PCM, which is strongly influenced by the natural convection takes an effective thermal conductivity $\kappa_{eff,\,L}$ into account. The effects of natural convection and close contact melting are modelled in the liquid PCM phase. The effective thermal conductivity is evaluated from experiments and detailed numerical models. During melting of PCM, it is well known that the influence of convection rises along with the increase of melting column. Different correlations were developed to describe the natural convection in melting column. A general formulation of the effective thermal conductivity in fluids with natural convection is written as [196]

$$\kappa_{eff} = \kappa_F \cdot \mathrm{Nu}. \tag{5.71}$$

Here, κ_F is the thermal conductivity of fluid and Nu is the Nusselt number. The effective thermal conductivity for the natural convection in a PCM melting column defined by Farid et al. [197] can be written as follows

$$\kappa_{eff,\,L} = \kappa_L \cdot \left[c\,\mathrm{Ra}_L^n \left(\frac{\delta}{l}\right)\right] \tag{5.72}$$

where κ_L is thermal conductivity of liquid PCM phase, l is the characteristic length, δ is the length of melting column, c, n are adaptable constants for melting conditions and Ra_L is the Rayleigh number, which is defined as

$$\mathrm{Ra}_L = g\beta c_L \left(\frac{\Delta T l^3}{\mu \kappa_L}\right). \tag{5.73}$$

Here, g is the value of the gravitational acceleration, ΔT is the difference between ambient and solid surface temperature $(T_w - T_m)$ and μ is the dynamic viscosity of the liquid PCM. However, the correlation as in Eq. 5.72 was only validated for a single heated wall [197]. Therefore in the present work, the correlation is adapted for a volumetric melting of PCM as in Eq. 5.74

$$\kappa_{eff,\,L} = \kappa_L \cdot c \cdot \left[\mathrm{Ra}_L \cdot V(\gamma_L)\right]^n, \tag{5.74}$$

where $V(\gamma_L)$ is the volume of melted PCM portion in the computational domain. From Eq. 5.72 and Eq. 5.71, it can be understood that the Nusselt number can estimated in the form of the Rayleigh number, which can be denoted in the form

$$\mathrm{Nu} = c\,\mathrm{Ra}_L^n. \tag{5.75}$$

The constants c and n are investigated and adapted for each melting condition. They are adapted for every melting condition for physical effects like natural convection, close contact melting and enhanced heat transfer. For a definite operating condition, the role of these physical effects on the heat transfer inside the PCM capsule are adapted. Therefore, the constants c and n vary depending on the temperature differences, geometrical configuration and PCM material properties. A summary of empirical correlations for different melting conditions are shown in Tab. 5.1.

Table 5.1: Estimation of the effective thermal conductivity of natural convection in geometries [197].

Correlation	Convection range	Melting set-up	Ref.
$\mathrm{Nu} = 0.159\mathrm{Ra}_L^{0.34}$	$3 \times 10^2 < \mathrm{Ra}_L < 3 \times 10^6$	Horizontal cylinder	[198]
$\mathrm{Nu} = 0.25\mathrm{Ra}_L^{0.25}$	$10^3 < \mathrm{Ra}_L < 4 \times 10^6$	Rectangular enclosure	[199]
or			
$\mathrm{Nu} = 0.072\mathrm{Ra}_L^{0.3333}$	$10^3 < \mathrm{Ra}_L < 4 \times 10^6$	Rectangular enclosure	[199]
$\mathrm{Nu} = 0.10\mathrm{Ra}_L^{0.25}$	$10^4 < \mathrm{Ra}_L < 10^7$	Vertical plane wall	[200]
$\mathrm{Nu} = 0.10\mathrm{Ra}_L^{0.25}$	$2 \times 10^3 < \mathrm{Ra}_L < 10^7$	Vertical cylinder	[197]
$\mathrm{Nu} = 0.28\mathrm{Ra}_L^{0.25}$	$10^9 < \mathrm{Ra}_L < 2.5 \times 10^{10}$	Vertical cylinder	[197]

Solution algorithm

Solving the melting of solid PCM with the effective thermal conductivity model is relatively effortless. Here, just a diffusive energy equation is solved to compute the melting of solid PCM. No effective thermal conductivity is necessary to model the solidification of liquid PCM. This model is capable to compute the melting and solidification within a low computational time. Inside a time step,

- the energy equation as in Eq. 5.70 is solved by employing the correlation of thermal conductivity described in Eq. 5.69.
- For solving the non-linear enthalpy-temperature the optimum approach is employed. The apparent heat capacity c^A is implemented in the energy equation as in Eq. 5.31.
- The enthalpy is updated as in Eq. 5.61 after solving the energy equation.
- The temperature is corrected from the updated enthalpy as described in Eq. 5.36.
- The liquid fraction is calculated using Eq. 5.37, which is responsible to estimate the natural convection and close contact melting in the computational domain.
- The convergence of the energy equation is thereby characterized with Eq. 5.38.

A brief overview of the simplification process is depcited in the form of flow chart in Fig. 5.6. The steps involved in solving the multiphase numerical model, the simplified convective model and the simplified effective conductive model for solid-liquid phase change are shown here.

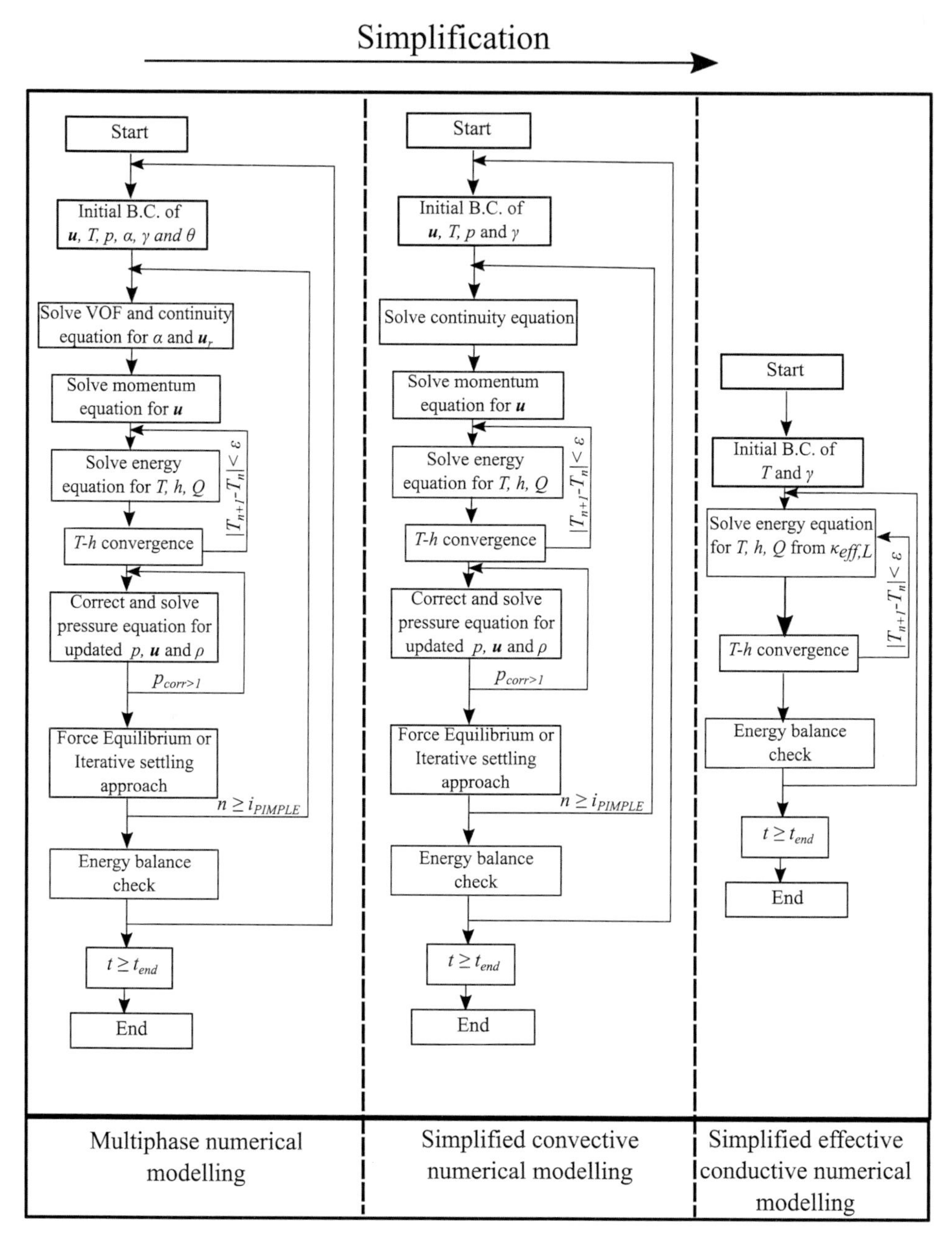

Figure 5.6: Flow diagram of different numerical models developed to solve solid-liquid phase change in a PCM capsule.

5.2 Modelling heat transfer in a latent heat thermal storage unit

Modelling solid-liquid phase change inside a single PCM capsule is not sufficient to optimize thermal performance of a thermal storage unit. Rather, a numerical model representing fluid flow and heat transfer of HTF along with solid-liquid phase change in PCM capsule is necessary. But, such coupled models are computationally expensive for a large thermal storage unit. Therefore in the present work, two different models are developed to represent the charging and discharging of a LHTES unit at different length scales. For medium length scale, a detailed conjugate heat transfer (CHT) model represents the interaction between PCM capsules and HTF in a storage unit. Due to its enormity, the conjugate heat transfer model can only compute the interaction of very few capsules with HTF in a acceptable time. Therefore in this work, the conjugate heat transfer model is only employed to compute heat transfer between three PCM capsules and HTF.

Alongside, a volume averaged local thermal non-equilibrium (LTNE) model is developed to compute thermal charging and discharging of a large scale LHTES unit. Here, fluid flow and heat transfer is solved for HTF and volume averaged effective thermal conductivity method is solved for PCM capsules.

5.2.1 Conjugate heat transfer (CHT) model

The developed CHT model represents the solid-liquid phase change inside PCM capsules and the convective heat transfer in HTF as shown in Fig. 5.7. The numerical model includes the following physical phenomena in a thermal storage unit:

- Heat transfer fluid (HTF) – Fluid and heat flow, heat transfer and heat losses,
- Phase change material (PCM) – Solid-liquid phase change from the heat transfer between PCM and capsule wall along with the settling of solid PCM during melting of PCM,
- PCM capsule wall – Heat transfer from HTF to capsule wall and capsule wall to PCM including thermal conductivity of the capsule wall.

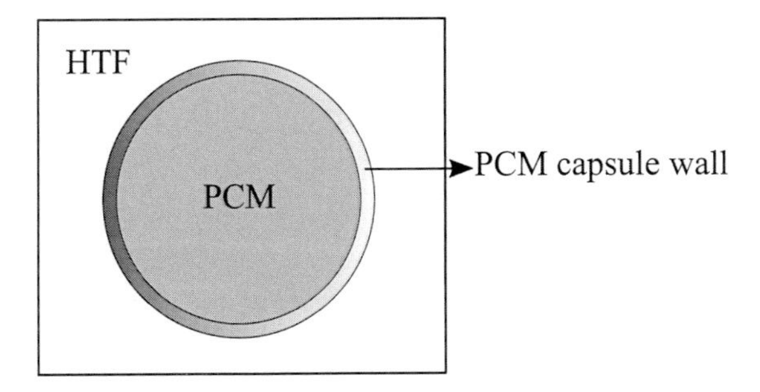

Figure 5.7: Schematic representation of the interaction between the PCM capsules and HTF in a LHTES unit.

For this, different set of governing equations are solved on distinctive grids. The fluid flow and heat transfer is solved for HTF. For the PCM, solid-liquid phase change with liquid PCM natural convection and solid PCM settling is solved. A diffusive energy equation is solved for the solid capsule wall. This thermally coupled fluid-structure model employ different computational domains to solve these physical phenomena. The interaction between different domains is modelled by enabling the contact of grid interfaces.

Fluid domain : Heat transfer fluid (HTF)

In general, the fluid responsible for charging and discharging of thermal storage unit has temperature dependent physical properties due to its operating temperature range. But for simplification only temperature dependent density is taken into account for HTF. Other material properties like viscosity, heat capacity and thermal conductivity are represented with constant values. For a temperature dependent density, mass is conserved with governing equations written as

$$\frac{\partial \rho_F}{\partial t} + \nabla \cdot (\rho_F \mathbf{u}_F) = 0. \tag{5.76}$$

Here, ρ_F is the density of HTF and $\mathbf{u}_F$ being the velocity of HTF. The density change is implemented through the thermo-physical library of OpenFOAM. The momentum in the HTF domain is conserved by Eq. 5.77

$$\frac{\partial \rho_F \mathbf{u}_F}{\partial t} + \nabla \cdot (\rho_F \mathbf{u}_F \mathbf{u}_F) = -\nabla p_d - \mathbf{g} \cdot \mathbf{x} \nabla \rho_F + \mu_F \Delta \mathbf{u}_F, \tag{5.77}$$

where μ_F is the corresponding dynamic viscosity of HTF. The energy within the HTF is conserved with

$$\frac{\partial \rho_F h}{\partial t} + \nabla \cdot (\rho_F \mathbf{u}_F h) = \nabla \cdot (\kappa_F \nabla T) + \dot{Q}_{FW} + \dot{Q}_L \tag{5.78}$$

where enthalpy in the HTF is represented by h, κ_F is the thermal conductivity of HTF, $\dot{Q}_{FW}$ is the heat out- or inflow into the HTF through the interface of capsule wall and $\dot{Q}_L$ are the heat losses in HTF. The heat transfer between the HTF and the PCM capsule wall is modelled using a film heat transfer approach where

$$\dot{Q}_{FW} = C_{ht} \, A_{FW} \left[T_F - T_{W,O} \right] = \frac{1}{\Delta V} \int\limits_{A_{FW}} \mathbf{n} \cdot \kappa_F \nabla T_F \, dA_{FW} = \frac{1}{\Delta V} \int\limits_{A_{FW}} \mathbf{n} \cdot \kappa_W \nabla T_{W,O} \, dA_{FW}. \tag{5.79}$$

With the present numerical model, a local heat transfer coefficient between HTF and PCM capsule wall C_{ht} is not required to compute the heat flow between PCM capsules and HTF $\dot{Q}_{FW}$. Instead, a temperature continuity is implemented at the interface. Where, κ_W is the thermal conductivity of capsule wall and A_{FW} is the contact surface area between HTF and capsule wall. The direction of heat flow between the domains is represented by normal vector $\mathbf{n}$ which points the direction of heat flow.

In Eq. 5.79, T_F and $T_{w,o}$ represent the temperatures of HTF and outer surface of capsule wall respectively. The interface between capsule wall and HTF acts as heat transfer surface, where computational domains of HTF and capsule wall interact as seen in Fig. 5.8. Heat transfer between the domains is occurred due to temperature gradient at the interface.

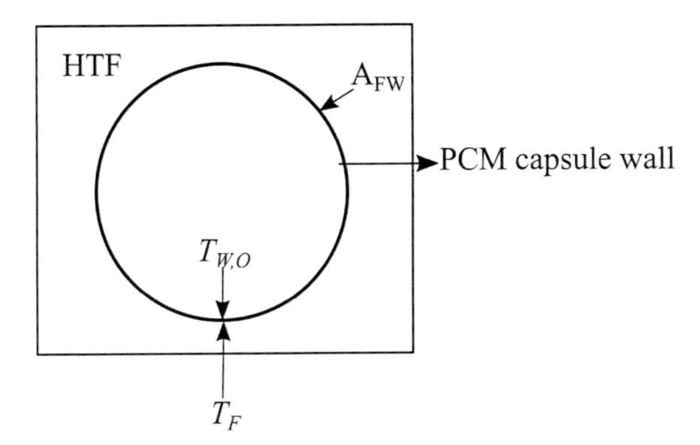

Figure 5.8: Computational domain for HTF including the heat transfer interface between the HTF and PCM capsule wall.

Solid domain : PCM-Capsule wall

During the charging and discharging of a thermal storage unit, PCM capsule wall experiences temperature gradient. In the present work PCM capsule wall is treated as an individual computational domain as seen in Fig. 5.9.

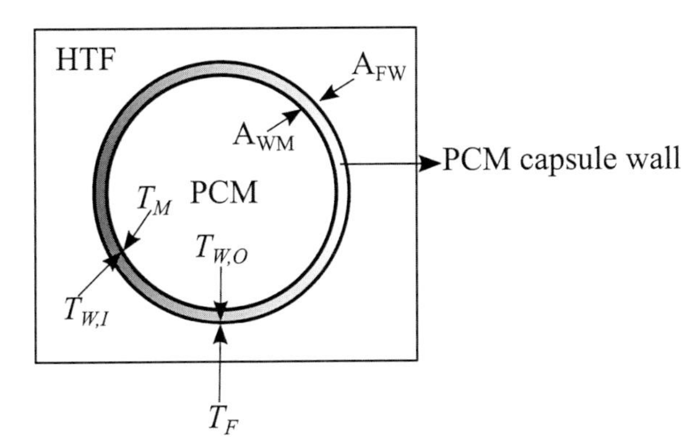

Figure 5.9: Computational domain for capsule wall sharing overlapping interfaces with PCM and HTF domains.

In solid capsule wall, the thermal transport is modelled with a simple diffusive equation. The conservation of energy in the capsule wall domain is written as

$$\frac{\partial \rho_W h}{\partial t} = \nabla \cdot (\kappa_W \nabla T) + \dot{Q}_{FW} + \dot{Q}_{WM}. \tag{5.80}$$

Here, ρ_W is the density of the capsule wall and κ_W being its thermal conductivity. Heat flow between the capsule wall and HTF $\dot{Q}_{FW}$ is shown in Eq. 5.80. Similarly, the heat flow between capsule wall and PCM $\dot{Q}_{WM}$ is shown in Eq. 5.81.

$$\dot{Q}_{WM} = \frac{1}{\Delta V} \int_{A_{WM}} \mathbf{n} \cdot \kappa_W \nabla T_{W,I} \; dA_{WM} = \frac{1}{\Delta V} \int_{A_{WM}} \mathbf{n} \cdot \kappa_M \nabla T_M \; dA_{WM}. \tag{5.81}$$

In Eq. 5.81, $T_{W,I}$ is the temperature of inner capsule wall surface and T_M is the temperature of PCM. The inner surface area of capsule wall is represented by A_{WM} with κ_M being the thermal conductivity of PCM phase adjacent to wall. Computational domain of capsule wall bridges the heat flow between PCM and HTF in a thermal storage unit.

Multiphase domain (solid-liquid): Phase change material (PCM)

For detailed modelling of heat transfer in a thermal storage unit, solid-liquid phase change is accounted in the PCM domain as seen in Fig. 5.10.

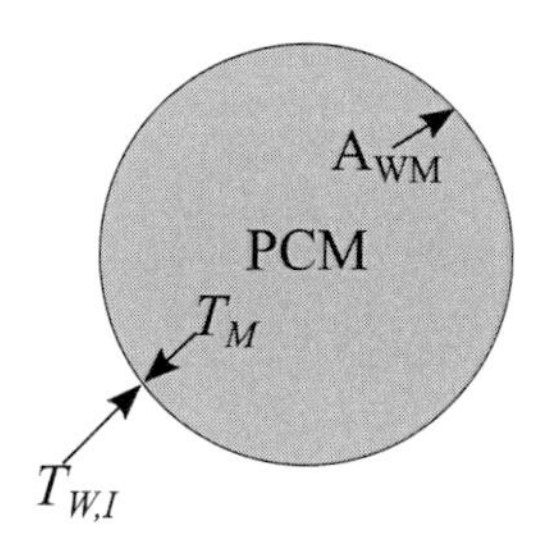

Figure 5.10: Pictorial depiction of computational domain employed to represent the solid-liquid phase change of PCM.

Therefore, the set of governing equations for solid-liquid phase change are solved in PCM domain. Conservative equations of mass, momentum and energy for incompressible flow are employed to depict the solid-liquid phase change. Governing equations are incorporated with simplified convective model presented in this work. Mass in the solid-liquid multiphase mixture of PCM domain is conserved by

$$\nabla \cdot \mathbf{u}_M = 0. \tag{5.82}$$

Here, $\mathbf{u}_M$ is the velocity in PCM domain. For a solid-liquid phase change, the momentum in PCM domain is conserved by

$$\frac{\partial \rho_M \mathbf{u}_M}{\partial t} + \nabla \cdot (\rho_M \mathbf{u}_M \mathbf{u}_M) = -\left(\nabla p_d\right)_M - \mathbf{g} \cdot \mathbf{x} \nabla \rho_{i,M} + \nabla \cdot (\mu_M \nabla \mathbf{u}_M). \tag{5.83}$$

The density from Bousinessq approximation can be formulated as

$$\rho_{i,M} = \gamma_{S,M} \rho_{S,M} + \gamma_{L,M} \rho_{L,M} \cdot (1 - \beta_M \Delta T_M). \tag{5.84}$$

In Eq. 5.84, $\gamma_{L,M}$ is the volume fraction of liquid PCM in PCM domain. The volume fraction of solid PCM is $\gamma_{S,M}$, which can be written as

$$\gamma_{S,M} = 1 - \gamma_{L,M}. \tag{5.85}$$

In Eq. 5.83, ρ_M is the constant density of PCM phase. The variable solid-liquid mixture density $\rho_{i,M}$ is modelled using Bousinessq approximation, where $\rho_{S,M}$ is the density of solid PCM and $\rho_{L,M}$ is the liquid PCM density. Volumetric expansion coefficient in the liquid phase of PCM is represented by β_M and ΔT_M is the corresponding temperature difference. In Eq. 5.83, $(\nabla p_d)_M$ is the pressure gradient in computational domain and μ_M is the viscosity of PCM defined as in Eq. 5.58.
The energy equation can be formulated as in Eq. 3.10, which is written as

$$\frac{D\rho_M h}{Dt} = c_M^A \frac{\partial \rho_M T}{\partial t} + c_M^A \nabla \cdot (\rho_M \mathbf{u}_M T) = \nabla \cdot (\kappa_M \nabla T) + \dot{Q}_{WM}. \tag{5.86}$$

The additional heat transfer source term $\dot{Q}_{WM}$ characterizes the heat transfer between PCM and the capsule wall as described in the Eq. 5.81. In Eq. 5.86, the apparent heat capacity of PCM c_M^A is defined as

$$c_M^A = \left[\gamma_{S,M} c_{S,M} + \gamma_{L,M} c_{L,M}\right] + \frac{L}{\sqrt{2\pi}\sigma_m} \cdot e^{-0.5\left[\frac{T - T_m}{\sigma_m}\right]^2}. \tag{5.87}$$

Here, $c_{S,M}$ is the specific heat capacity of solid PCM and $c_{L,M}$ is the specific heat capacity of liquid PCM. The energy equation as in Eq. 5.86 is solved using optimum approach. In optimum approach, energy equation in the discretized form can be represented as in Eq. 5.34. After solving the energy equation, enthalpy is updated by

$$h^{m+1} = h^m + \omega \cdot c_M^A \cdot (T^{m+1} - T^m). \tag{5.88}$$

For enthalpy-temperature consistency, temperature is corrected from updated enthalpy similar to Eq. 5.62. The liquid fraction of PCM $\gamma_{L,M}$ is calculated from the corrected temperature as in Eq. 5.89.

$$\gamma_L = 0.5 + 0.5\text{erf}\left[\frac{T - T_m}{\sqrt{2}\sigma_m}\right] \approx 0.5 + 0.5\tanh\left[\frac{T - T_m}{\sigma_m}\right]. \tag{5.89}$$

The convergence of the energy equation is governed by a convergence criterion ε_{conv} as

$$|T_p^{m+1} - T_p^m| < \varepsilon_{conv}. \tag{5.90}$$

The iteration loop is executed till the convergence is achieved.

Close contact melting

The explicit coupling of different computational domains like fluid, solid and multiphase system demands operating at small time steps. Therefore, an implicit method is employed to represent the solid settling inside a PCM capsule. Here, the selected low time step drives the convergence of pressure and velocity in the PCM domain influencing solid PCM settling. The density difference between the solid and liquid phases as described in Eq. 5.83 and Eq. 5.84 is responsible for solid PCM settling. Here, in the momentum equation

$$\nabla \cdot (\mathbf{u}_M)_P^* = \nabla \cdot \left[\frac{(\nabla p)_P}{\mathbf{A}_P} \right] \tag{5.91}$$

is solved. Where, the velocity at the computing node P is $(\mathbf{u}_M)_P$.

Solution algorithm

To compute the heat transfer between PCM capsules and HTF, the CHT model is employed. As shown in Fig. 5.11, the heat transfer between three PCM capsules and HTF is studied in the present work.

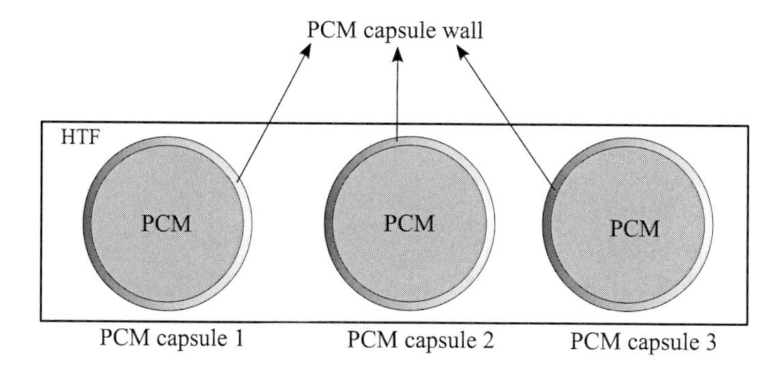

Figure 5.11: Numerical set-up to compute conjugate heat transfer between PCM capsules and HTF.

For that, the conservation of following physical quantities is necessary:

- Heat transfer fluid (HTF) - Conservation of mass, momentum and energy
- PCM capsule wall 1 - Conservation of energy
- PCM capsule wall 2 - Conservation of energy
- PCM capsule wall 3 - Conservation of energy
- PCM in capsule 1 - Conservation of multiphase mass, momentum and energy
- PCM in capsule 2 - Conservation of multiphase mass, momentum and energy
- PCM in capsule 3 - Conservation of multiphase mass, momentum and energy

The steps involved to solve conjugate heat transfer in a thermal storage unit are represented schematically as shown in Fig. 5.12.

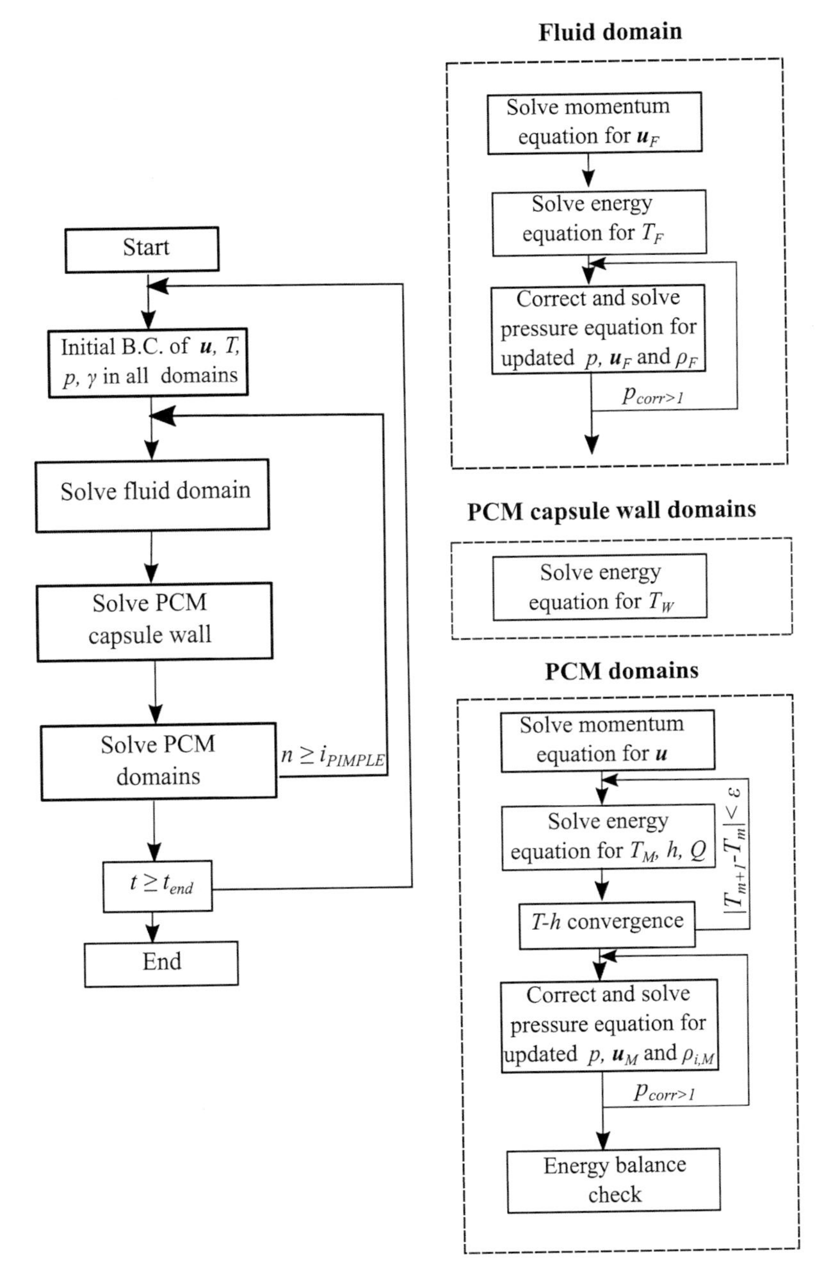

Figure 5.12: Flow diagram showing the process sequence developed to compute conjugate heat transfer between PCM capsules and HTF.

5.2.2 Local thermal non-equilibrium (LTNE) model for a thermal storage unit

Computing heat transfer and fluid flow during charging and discharging of a large scale thermal storage unit is not possible within an acceptable time using the developed CHT model. Therefore, a simplified approach to compute the charging and discharging of a thermal storage unit is introduced. This numerical model employs a homogenised approach to represent the heterogeneous distribution of PCM capsules in a thermal storage unit. Inside a thermal storage unit, the fluid-solid system is homogenized as porous media with a volume averaging approach. The porosity ε_F defined in a thermal storage unit is

$$\varepsilon_F = \frac{V_F}{V_B} \tag{5.92}$$

where V_F is the volume of HTF in a storage unit and V_B is the volume of packed bed in a thermal storage unit. Similarly, the volume fraction of PCM capsules in a packed bed of the storage unit ε_M can be defined as

$$\varepsilon_M = 1 - \varepsilon_F = \frac{V_M}{V_B}. \tag{5.93}$$

Here, V_M is the total volume occupied by PCM capsules inside a storage unit. The volume is homogenized over a REV, which behaves as a representative element of the material distribution in the thermal storage unit. A representative quantity of the heterogeneous multiphase mixture of PCM capsules and HTF in the packed bed of a thermal storage unit is shown in the Fig. 5.13. The porosity of a unit cell is representative of the total porosity distribution of thermal storage bed. During thermal charging, the PCM capsules in a thermal storage unit have relatively lower temperature than HTF. Similarly during thermal discharging, the PCM capsules stay at higher temperature than HTF. To represent the heat transfer between the PCM capsules and HTF, thermal non-equilibrium between them is implemented in the numerical model. Thereby, a local thermal non-equilibrium (LTNE) model is adapted to represent the heat transfer between PCM capsules and HTF in a thermal storage unit. Due to different material properties of PCM capsules and HTF, they are treated as two different continuum interacting with each other. Here, the LTNE model is implemented on two computations domains to represent the multiphase system in macro-encapsulated LHTES unit. The fluid flow and heat transfer of the HTF is modelled on a computational domain, which is treated as a porous continuum. PCM capsules are treated on a different computational domain as another continuum. These two computational domains are coupled with a volumetric heat transfer condition characterized by the heat flow between PCM capsules and HTF. In a thermal storage, HTF and PCM capsules may have different volumes. Therefore, they are treated as identical overlapping computational domains with different porosities. For fluid domain, the porosity in Eq. 5.92 is employed. Thereby, the volume averaged governing equations of mass, momentum and energy are employed to represent the fluid flow and heat transfer in HTF. An intrinsic averaged physical quantity $\langle\phi\rangle_i$ for a HTF domain is defined as

$$\langle\phi\rangle_i = \frac{1}{V_F}\int_{V_F} \phi dV. \tag{5.94}$$

The superficial averaged physical quantity $\langle\phi\rangle_s$ for a HTF domain is defined as

$$\langle\phi\rangle_s = \frac{1}{V}\int_{V_F} \phi dV = \varepsilon_F \cdot \langle\phi\rangle_i. \tag{5.95}$$

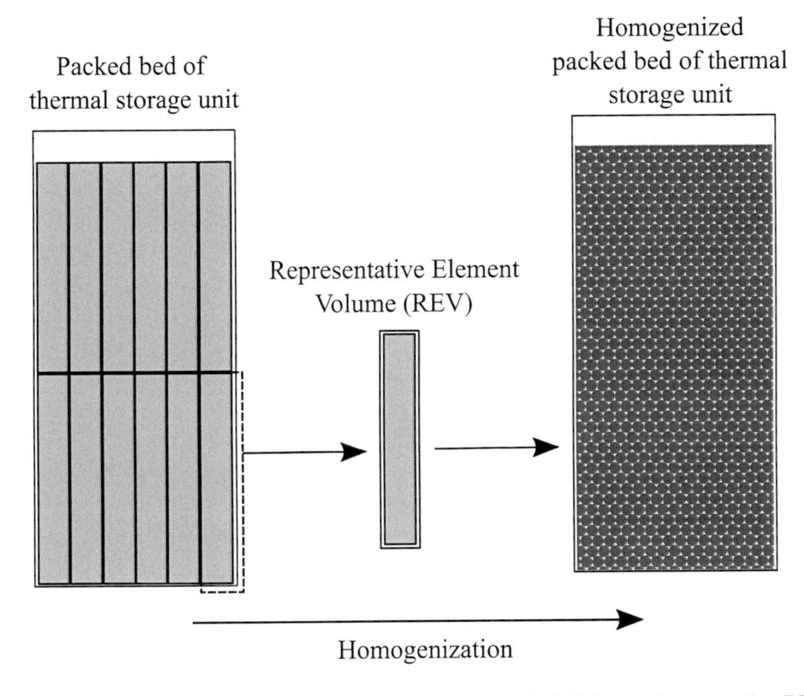

Figure 5.13: Homogenization of the packed bed with a single PCM capsule selected as REV.

Fluid domain: Heat transfer fluid (HTF)

The governing equations representing the physical process in HTF are solved on the fluid computational domain. A homogenized porous approach with the LTNE model offers a volume reduction of total computational domain. In this case , the active volume of the computational domain is equal to the real volume of HTF inside a thermal storage unit. The intrinsic averaging of a physical quantity in fluid domain can be represented as the local volume averaging quantity, which is represented as [179, 201]

$$\langle\phi\rangle = \langle\phi\rangle_i = \frac{1}{V_F}\int_{V_f} \phi dV. \tag{5.96}$$

Following Amiri and Vafai [179], the local volume averaged mass conservation of the HTF is governed by

$$\nabla \cdot \langle\mathbf{u}\rangle = 0. \tag{5.97}$$

Similarly, the local volume averaged momentum equation of HTF is defined as

$$\frac{\rho_F}{\varepsilon_F}\left[\frac{\partial\langle\mathbf{u}\rangle}{\partial t}+\langle(\mathbf{u}\cdot\nabla)\mathbf{u}\rangle\right]=-\nabla\langle p\rangle+\frac{\mu}{\varepsilon_F}\nabla^2\langle\mathbf{u}\rangle-\frac{\mu}{K}\langle\mathbf{u}\rangle-\frac{\rho F\varepsilon_F}{\sqrt{K}}\left[\langle\mathbf{u}\rangle\cdot\langle\mathbf{u}\rangle\right]+\rho_F\cdot\mathbf{g}. \quad (5.98)$$

For anisotropic porous structure the momentum equation is derived to

$$\frac{\rho_F}{\varepsilon_F}\left[\frac{\partial\langle\mathbf{u}\rangle}{\partial t}+\langle(\mathbf{u}\cdot\nabla)\mathbf{u}\rangle\right]=-\nabla\langle p\rangle+\frac{\mu}{\varepsilon_F}\nabla^2\langle\mathbf{u}\rangle-\mathbf{D}_{ij}\,\mu\langle\mathbf{u}\rangle-0.5\rho\mathbf{F}_{ij}\,|\mathbf{u}|\langle\mathbf{u}\rangle+\rho_F\cdot\mathbf{g}. \quad (5.99)$$

Here, D_{ij} is the Darcy tensor which defines the linear momentum resistance in the packed bed. The non-linear momentum resistance in the packed bed is accounted by $\mathbf{F}_{ij}$. In Eq. 5.98 and Eq. 5.99, ρ_F is the density of HTF and $\langle p\rangle$ is the volume averaged pressure, as measured in pressure gauge. The resistance offered by PCM capsules in packed bed can be implemented in the numerical model with empirical correlations. For spherical PCM capsules, the momentum resistance offered in a thermal storage unit can be implemented by a modified Ergun equation introduced by Brauer [202] as

$$\nabla p=160\frac{(1-\varepsilon)^2}{\varepsilon^3}\frac{\mu_F u_s}{d_p^2}+3.1\frac{(1-\varepsilon)}{\varepsilon}\frac{\rho_F u_s^2}{d_p^2}\left[\frac{\mu_F(1-\varepsilon)}{\rho_F\mu_F d_p}\right]^{0.1}. \quad (5.100)$$

However, for a vertical cylindrical distribution such a simple correlation does not exist. Therefore to estimate the pressure drop in a packed bed with vertically positioned cylindrical PCM capsules, an equation is drawn from the pressure loss correlation for shell and tube heat exchangers. Accordingly, the pressure drop for a packed bed with vertical cylinders can be defined as [137]

$$\Delta p=\nabla p\cdot L=\xi\left(\frac{L}{d_h}\right)\left(\frac{\rho_F w_m^2}{2}\right). \quad (5.101)$$

Here, L is the length of packed bed, w_m is the mean velocity through packed bed defined as

$$w_m=\frac{\dot{V}}{A_m}. \quad (5.102)$$

In Eq. 5.101, the hydraulic diameter d_h can be defined as

$$d_h=4\left[\frac{A_B}{C}\right]. \quad (5.103)$$

Here, A_B is the active area of packed bed useful for flow and C is the total circumference of PCM capsules and packed bed container. The drag coefficient ξ is defined differently based on the mean Reynolds number, for laminar flow with $Re_m < 2300$ the drag coefficient is defined as

$$\xi=\frac{64}{Re_m}. \quad (5.104)$$

For turbulent flows with mean Reynolds number Re_m ranging between 3000 and 10^5, the drag coefficient is defined as

$$\xi = \frac{0.3164}{Re_m^{0.25}}. \tag{5.105}$$

The permeability can be defined from the pressure drop Δp defined as

$$K = \frac{\mu_F L}{\Delta p} \cdot u_s. \tag{5.106}$$

Thereby, the Darcy term as in Eq. 5.99 can be defined as

$$D_{ii} = \frac{1}{K_i} \tag{5.107}$$

The Forchhemier term is defined for inertial driven flows with Reynolds numbers higher than 10. Implementing the pressure segregation from Eq. 5.19, equation for the conservation of momentum in HTF becomes

$$\frac{\rho_F}{\varepsilon_F}\left[\frac{\partial\langle\mathbf{u}\rangle}{\partial t} + \langle(\mathbf{u}\cdot\nabla)\mathbf{u}\rangle\right] = -\nabla\langle p_d\rangle - \nabla\rho_{i,F}\,\mathbf{g}\cdot\mathbf{x} + \frac{\mu}{\varepsilon_F}\nabla^2\langle\mathbf{u}\rangle - \mathbf{D}_{ij}\,\mu\langle\mathbf{u}\rangle - 0.5\rho\mathbf{F}_{ij}\,|\mathbf{u}|\langle\mathbf{u}\rangle. \tag{5.108}$$

The density change in HTF is accounted with Bousinessq approximation, where

$$\rho_{i,F} = \rho_F\left(1 - \beta_F\cdot\Delta T_F\right). \tag{5.109}$$

Alongside, the conservation of energy is also modelled using a homogenized approach. The energy conservation equation is modelled with a continuous solid phase approach. For this, the conservation of energy in HTF is formulated as

$$\rho_F\left[c_F\frac{\partial\varepsilon_F\langle T_F\rangle}{\partial t} + \nabla\cdot\left(c_F\langle u\rangle\langle T_F\rangle\right)\right] = \nabla\cdot\left(\kappa_{F,Eff}\nabla\langle T_F\rangle\right) + U\cdot A\cdot\left(\langle T_S\rangle - \langle T_F\rangle\right). \tag{5.110}$$

Here, c_F is the specific heat capacity of HTF, $\langle T_F\rangle$ is the intrinsic phase average temperature of HTF and $\kappa_{F,Eff}$ is the effective thermal conductivity defined as

$$\kappa_{F,Eff} = \varepsilon_F\,\kappa_F. \tag{5.111}$$

In Eq. 5.110, the specific heat transfer surface area A is defined as

$$A = \varepsilon_S\cdot\left(\frac{A_c}{V_c}\right) \tag{5.112}$$

which can be explained as the heat transfer surface area A_c accessible per capsule volume V_c. Whereas, the overall heat transfer coefficient U is computed from the set of different thermal resistances involved in heat transfer between HTF and PCM capsules.

Solid domain: PCM capsule bed

It is well known that the fluid domain due to its physical form is generally treated as a continuum. But in homogenized approach, a packed bed with tight packing can also be treated as continuum. The homogenized volume of PCM capsules in the packed bed is treated on a different computational domain. For this, the volume fraction of PCM capsules ε_S is employed. Because, the total volume of thermal storage container is modelled as fluid and solid domains. In the present storage unit, cylindrical PCM capsules are stacked on one another as packed bed. For this, a continuous solid phase approach is employed to model the solid PCM capsules in a packed bed. The assumption of this approach lies in the consideration of continuous PCM capsule distribution in a storage unit. The PCM capsules distribution in the storage unit is governed by Eq. 5.93. The conservation of energy in the solid phase is written as

$$\varepsilon_S c_S \frac{\partial \rho_S \langle T_S \rangle}{\partial t} = -U \cdot A \cdot (\langle T_S \rangle - \langle T_F \rangle) \tag{5.113}$$

where c_S is the effective heat capacity of PCM capsule computed through the weighted average of capsule wall and PCM. The weighted volume average of capsule wall in a PCM capsule is defined as χ_w, which is used to correspond the material properties of the PCM capsule. Thereby, the corresponding effective specific heat capacity of the PCM capsule can be defined as

$$c_S = \chi_w \cdot c_W + (1 - \chi_w) \cdot c_M^A, \tag{5.114}$$

where, c_W is the specific heat capacity of PCM wall and c_M^A is the apparent heat capacity of PCM. The specific heat capacity of PCM is implemented as in Eq. 5.87. Here, the optimum approach is employed to solve the energy equation. Similar to other properties, the density of PCM capsule is also calculated as a weighted average in Eq. 5.115.

$$\rho_S = \chi_w \cdot \rho_W + (1 - \chi_w) \cdot \rho_M \tag{5.115}$$

where ρ_W is the density of capsule wall and ρ_M is the density of PCM. For heat transfer process, the overall heat transfer coefficient U is calculated from the total thermal resistance, which is defined as in Eq. 5.116

$$R_T = \frac{1}{U \cdot A_c} = R_c + R_w + R_{PCM} \tag{5.116}$$

where R_T is the total thermal resistance offered for heat transfer between PCM capsules and HTF. Here, resistance from external convection around PCM capsule is defined as R_c, resistance through PCM capsule wall is R_w and resistance offered in PCM is R_{PCM}. These thermal resistances generally vary depending on the geometrical configuration of PCM capsule.

Thermal resistances: Vertical cylindrical storage capsules

Thermal resistance due to the convection of HTF around PCM capsules is denoted by R_c, which is mathematically represented as

$$R_c = \frac{1}{h_c \cdot A_c}. \tag{5.117}$$

Here, h_c is the heat transfer coefficient corresponding to heat transfer around PCM capsule and A_c is the surface area of PCM capsule. The mean heat transfer coefficient Nu_m in the packed bed is calculated with a correlation as in Eq. 5.118 [203].

$$\mathrm{Nu}_m = 4.08\left(1+0.0349\mathrm{Gz}_m^{1.46}\right)^{0.25}. \tag{5.118}$$

Here Gz_m is the mean Graetz number, which characterizes laminar flow in ducts written as

$$\mathrm{Gz} = \frac{\dot{m}c}{\kappa l} \tag{5.119}$$

where $\dot{m}$ is the mass flow rate of HTF, c is the specific heat capacity of HTF, κ is the HTF's thermal conductivity and l is the length of vertical cylinders in packed bed. Similarly, thermal resistance offered by the capsule wall R_w is defined as

$$R_w = \frac{\ln\left(\frac{r_o}{r_i}\right)}{2\pi l \kappa_w} \tag{5.120}$$

where r_o is the outer radius of PCM cylinder, r_i is the inner radius of PCM cylinder and κ_w is the thermal conductivity of cylinder.

Thermal resistance of PCM inside the capsule, R_{PCM} varies depending on the heat transfer during phase change. The thermal resistance is computed from the phase change volume fraction in a zero dimensional manner. The average value in of phase change volume fraction in the complete packed bed decides the mean thermal resistance of PCM. For an unsteady charging and discharging of storage unit, thermal resistance is accordingly modelled unsteady. Here, the thermal resistance of PCM is defined as

$$R_{PCM} = \frac{R^l_{PCM} \cdot R^r_{PCM}}{R^l_{PCM} + R^r_{PCM}}. \tag{5.121}$$

In Eq. 5.121, the total thermal resistance in PCM is the parallel combination of lateral and radial thermal resistances offered by the PCM phase change column. During solidification, it is well known that the solid column is formed at the capsule boundary and propagates to the center. But in melting, the fluid melt column starts at the capsule wall and the solid PCM settles down to capsule bottom due to buoyancy. This is discussed as close contact melting in the earlier sections of this work. Using LTNE approach it is difficult to directly model natural convection and close contact melting during a charging process. Therefore, a simplified approach is employed to account the close contact melting and natural convection inside a PCM capsule. Here, an effective thermal conductivity is employed to estimate the natural convection including close contact melting in a PCM capsule. Accordingly, the resistances offered in radial and lateral directions are shown in Fig. 5.14. Thermal resistance of PCM in lateral stretch from top to bottom is defined as

$$R^{l}_{PCM} = \frac{(l_{PCM} - l_{NP})}{\pi r_i^2 \kappa_{Eff}}. \tag{5.122}$$

Here, l_{PCM} is the length of PCM filled inside the cylinder, l_{NP} is the length of PCM with no phase change, r_i is the inner radius of cylindrical PCM capsule and κ_{Eff} is the effective thermal conductivity of PCM. Similarly, the thermal resistance offered by PCM in radial direction of the capsule is written as

$$R^{r}_{PCM} = \frac{\ln\left(\frac{r_i}{r_{NP}}\right)}{2\pi l_{PCM} \kappa_{Eff}}. \tag{5.123}$$

Here, r_{NP} is the radius of unchanged PCM phase. The dimensions of unchanged phase like length l_{NP} and radius r_{NP} are estimated using the solid and liquid volume fractions of PCM. From the calculated volume fraction for a phase the length and radius of the unchanged phase in the cylinder is calculated. This can be written as

$$l_{NP} = \gamma^{0.333} \cdot l_{PCM} \tag{5.124}$$

and

$$r_{NP} = \gamma^{0.333} \cdot r_i. \tag{5.125}$$

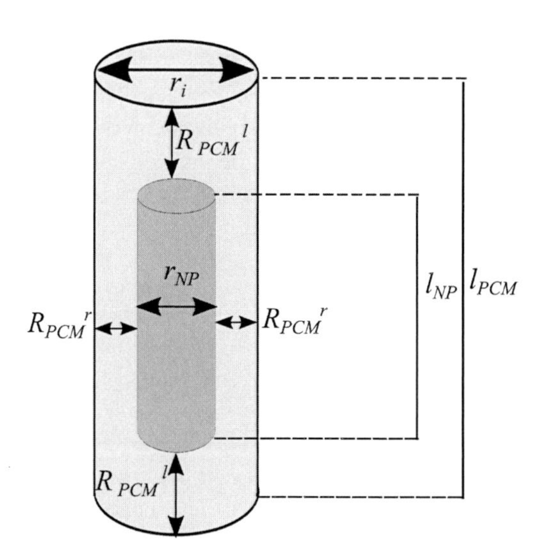

Figure 5.14: Thermal resistances offered by PCM to the heat transfer, due to the phase change column.

The total volume of transitioned phase inside the cylindrical domain is segregated into respective proportions, radially and laterally. The volume fraction of PCM, γ defined in Eq. 5.124 and Eq. 5.125 represents the liquid or solid PCM volume fraction depending on the case. For charging case, the volume fraction of PCM γ represents the liquid volume fraction of PCM, defined as

$$\gamma_L = 0.5 + 0.5 \tanh\left[4\left(\frac{T - T_m}{T_L - T_S}\right)\right]. \tag{5.126}$$

For discharging, the volume fraction of PCM γ denotes the solid volume fraction of PCM in the numerical model written as

$$\gamma_S = 1 - \gamma_L. \tag{5.127}$$

Thermal resistance offered by PCM for the heat exchange between PCM capsules and HTF increases with the growth of phase change column inside a capsule. During phase change the major heat exchange occurs between the solid-liquid interface of PCM and HTF. Therefore, the distance between the solid-liquid interface and HTF increases during phase change. This is implemented in the numerical model with a piecewise thermal resistance, where the thermal resistance offered by the melt column in the PCM is switched off till the volume fraction of phase change portion reaches 10% of the capsule inner volume. This is mathematically represented as

$$R_{PCM} = \begin{cases} \dfrac{R^l_{PCM} \cdot R^r_{PCM}}{R^l_{PCM} + R^r_{PCM}}, & \gamma \leq 0.9 \\ 0, & \gamma > 0.9. \end{cases} \tag{5.128}$$

To estimate the natural convection and close contact melting during melting, an effective thermal conductivity is employed. For solid PCM, this effective thermal conductivity κ_{Eff} is equal to the real thermal conductivity. Therefore, the effective thermal conductivity of solid PCM can be defined as

$$\kappa_{Eff} = \kappa_s. \tag{5.129}$$

The natural convection in liquid PCM increases with the melt column increment. Thereby, the effective thermal conductivity can be represented as in Eq. 5.71 and Eq. 5.74, which is written as

$$\kappa_{Eff} = \kappa_l \cdot \left[C \,\mathrm{Ra}^n \, V_{\alpha_l}\right]. \tag{5.130}$$

Here, κ_l is the thermal conductivity of liquid PCM, V_{α_l} is the volume of maximum melt of cylinders in packed bed, C and n are the geometrical set-up constants. Thus, overall heat transfer coefficient is calculated from different thermal resistances in the heat transfer process. Thereby, the total power in a LHTES unit is computed to estimate the power output and input. The power output of a thermal storage packed bed is implemented as

$$P = \dot{m}\, c_F \cdot (T_{out} - T_{in}) = \rho_F \, A_{out} \, u \, c_F \cdot (T_{out} - T_{in}). \tag{5.131}$$

Here, $\dot{m}$ is the mass flow rate of HTF, c_F is the specific heat capacity of HTF, T_{out} is the temperature of out-flowing HTF, T_{in} is the inlet temperature of HTF, A_{out} is the area of outlet region and u is the velocity magnitude of HTF.

Solution algorithm

Algorithm steps involved to solve the heat transfer between PCM capsule bed and HTF is shown in the following steps. To compute charging and discharging processes, the following steps are executed. In a time step,

- initially the pre-defined porosities of HTF and PCM capsule bed, temperature, pressure and velocity of the HTF and PCM capsules are defined.
- Later the set of mass, momentum and energy conservations for HTF are computed by solving Eq. 5.97, Eq. 5.108 and Eq. 5.110.
- The pressure of HTF is corrected with a PIMPLE correction from solved velocity and temperature.
- The overall heat transfer coefficient is thereby calculated by computing resistances as in Eq. 5.116, Eq. 5.117, Eq. 5.120 and Eq. 5.121.
- To calculate thermal resistance inside a PCM capsule, Eq. 5.124 and Eq. 5.125 are solved to obtain average value of phase change column in the packed bed.
- Based on the obtained dimensions of unchanged PCM phase of PCM inside a PCM capsule, PCM thermal resistance is obtained by solving Eq. 5.128. From the total thermal resistance, an overall heat transfer coefficient is computed.
- Later, the energy equation for the PCM packed bed is solved from the obtained overall heat transfer coefficient U alongside solcing HTF temperature T_F from Eq. 5.113.

Here, the initial boundary conditions vary between charging and discharging cases, where a higher temperature is implemented on fluid domain for charging and on solid domain for discharging. A schematic representation of the algorithm is shown in Fig. 5.15.

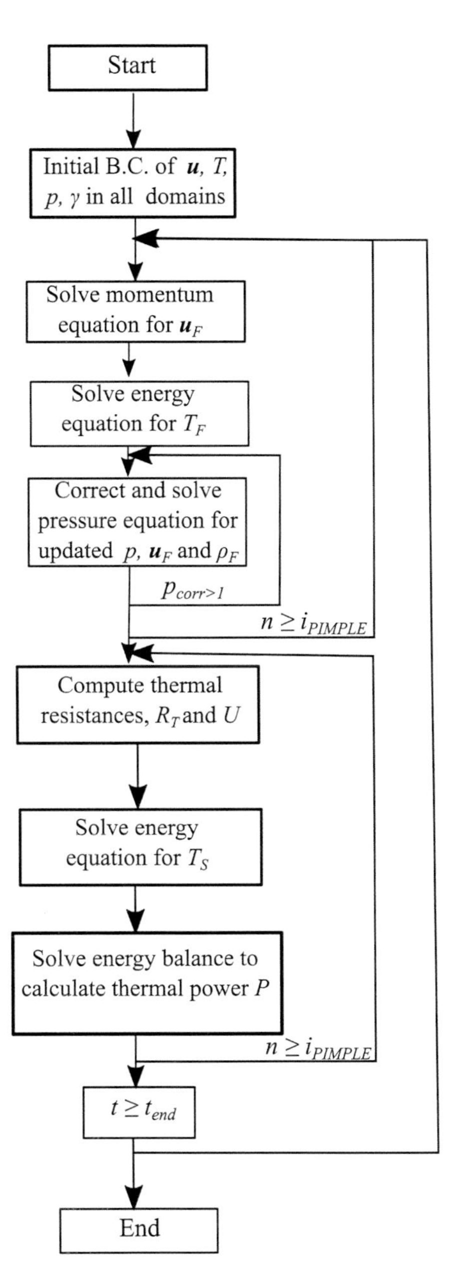

Figure 5.15: Algorithm steps developed to calculate the heat transfer and fluid flow in the latent heat thermal storage unit with the LTNE model.

6 Experimental methods

In the present section of work, experimental methods adapted to validate the numerical model will be explained. Different experimental set-ups were employed to represent the physical processes as in numerical simulations. An experimental set-up to validate the numerical model for solid-liquid phase change is adapted in the present work. Further, the conjugate heat transfer between the PCM capsules and HTF is studied experimentally with three cylindrical capsules immersed in a container. Here in addition to the heat transfer between the HTF and PCM capsules, melting of PCM in the cylindrical capsules is also studied. Additionally heat transfer induced during the charging and discharging of thermal storage unit is investigated on a laboratory scale latent heat thermal energy storage unit.

6.1 Optical investigation of phase change in a cubical cavity

Visual acquisition of physical processes to understand and investigate the details is often observed in different fields. Also, measurement techniques like particle image velocimetry (PIV) and laser interferometry have evolved into very sophisticated fields. Therefore, in the present work an optical measurement technique is employed to evaluate the solid-liquid phase change in PCM. However, this method is limited to evaluate the melting of solid PCM because during solidification the solid-liquid interface is hidden inside the opaque solid and cannot be analysed. Therefore, in the present work optical analysis is just confined to melting. Meanwhile, during melting the change of phase influenced by liquid convection and solid settling in the cavity makes it more complex than solidification. Additionally, the nature of convection and solid settling strongly depends on the layout of the experimental set-up. Thus in the case of melting an accurate methodology like the optical measurement is more relevant.

The experimental set-up for the optical investigation of PCM melting involves a cubical cavity inside which the solid to liquid phase change occurs. The cavity with dimensions 40 mm$\times$40 mm$\times$40 mm has an inlet and outlet, which are connected to the Lauda RK 8 KP thermostat through an insulated tube. The thermostat pumps the HTF around the cubical cavity with adjusted temperature. The temperature in the cubical cavity is therefore influenced by the HTF temperature. The top, bottom, left and right walls of the cavity are integrated as a single unit block made of copper. The front and back sides of the cavity are covered using two square transparent thermo-plastic discs to enable translucency in the cavity. Due to these plastic discs, the experimental set-up can only be operated at temperatures below 60 °C. The plastic discs mounted on the front and back walls of the cavity are coated with indium tin oxide, which acts as semi-conductor. These coated square discs are connected to a UNIWATT NG-310 power supply unit. The electric power supplied through the power supply unit is converted into thermal power on the coated square discs. However, the major heat flux into the cubical cavity is governed by the flow of HTF around the cubical cavity. Due to which the solid tends to melt from the side walls and remains fixed on the front and back walls resisting the settling. This could be solved, by inducing small heat flux on the front and back coated discs of the cubical cavity. The heat flux induced by the electric power supply unit effects in a negligible melting of front and back sides of PCM in the cubical cavity. The cavity can be filled with different proportions of PCM. In general, during melting the melted PCM needs more volume than solid inside the cavity due its lower density. The volumetric expansion of the PCM also builds pressure inside the cavity which may rupture the thermo-plastic walls. Therefore, two pressure outlets are arranged to compensate the pressure in the cavity by air out flow.

Additionally, a small LED light is placed behind the cubical cavity to illuminate the cubical cavity interior as seen in Fig. 6.1. As the solid phase of PCM is opaque, the light from the LED light only travels through the liquid phase of PCM in the cubical cavity. Hereby, the solid-liquid interface can identified to estimate the melted portion of PCM inside the cubical cavity.

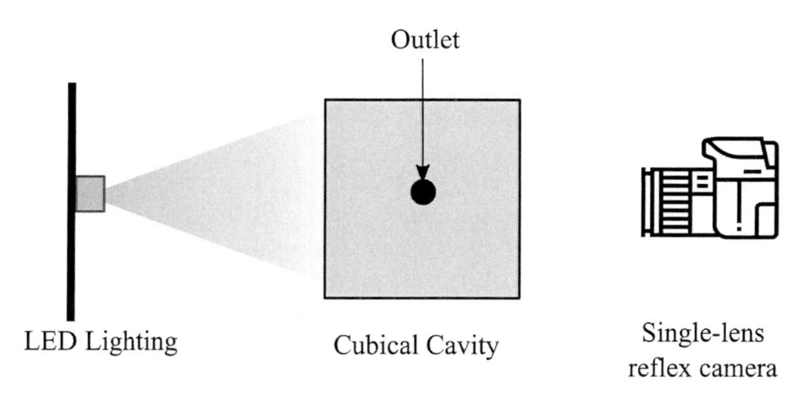

Figure 6.1: Side view of the experimental set-up depicting the incidence of light on the phase changing PCM in the cubical cavity, which is captured by the single-lens reflex camera [79, 204].

Two control valves regulate the flow of HTF around the cubical cavity, to adapt the desired temperature at the boundary. An additional bypass tube regulates the flow from entering the inlet and exiting the outlet. By employing the bypass mode, the flow of the HTF around the cubical cavity is interrupted. Therefore, in the bypass mode the fluid around the cubical cavity remains uninfluenced by the thermostat. In the bypass mode, the fluid in thermostat can be preheated till a required temperature. Later by turning off the bypass mode, HTF is allowed to flow around the cavity with a preheated temperature. The HTF enters the cavity housing through inlet and leaves the cavity housing from outlet. From the heat flux around the side walls the PCM inside the cubical cavity melts and solid PCM settles down due to its larger density. This complete physical process of melting and settling of the PCM inside the cubical cavity is captured using a single-lens reflex camera, Panasonic Lumix GH4 with a lens H-FS-FS12060. The camera is placed at the distance of 600 mm from the front surface of the cubical cavity as shown in Fig. 6.2. The photographs captured every minute depict the melting and settling of solid PCM inside the cubical cavity.

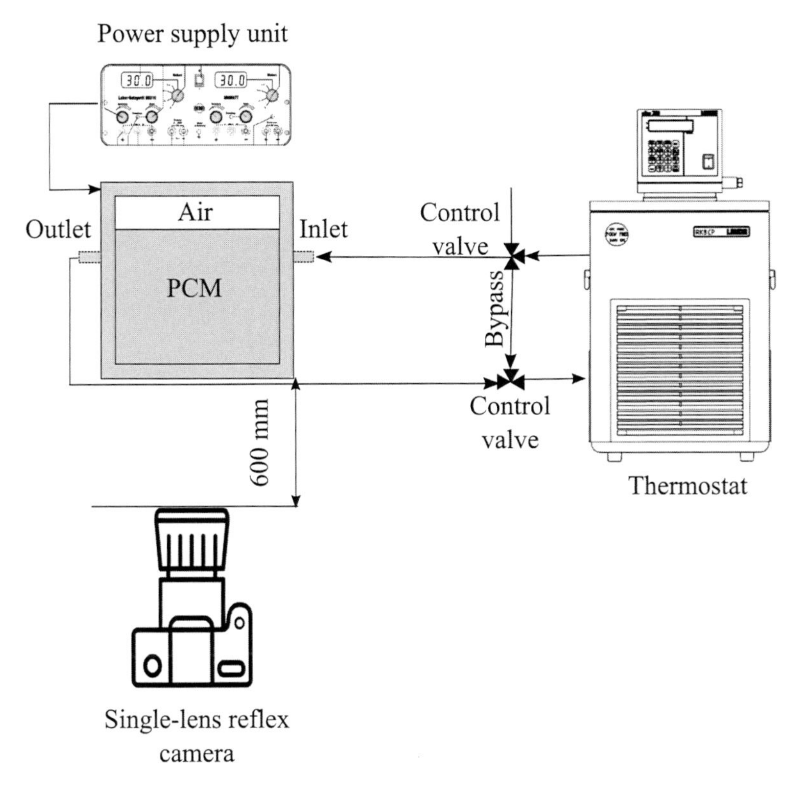

Figure 6.2: 2D view of experimental set-up showing the cubical cavity filled with PCM and air which is connected to a Thermostat and Power supply unit [205–207].

Later, these 2D photographs are analysed to calculate the melted portion of the solid PCM. Some inaccuracies in the experiments are influenced by the heat fluxes from front and back sides. However, these heat fluxes are very low in comparison to the total heat flux from the rest of walls. Generally, void spaces inside the solid PCM may also influence the melt rate of PCM. These void spaces are generally formed from the dissolved air phase in the liquid PCM and are locked inside the solid after solidification of liquid PCM. During melting, this dissolved air is freed in the form of bubbles which may interrupt the close contact melting. Therefore, degasification was employed to remove the dissolved air in the liquid PCM. For this, a small vacuum pump is used to degasify the dissolved air in the liquid PCM. Later, the degasified liquid PCM is filled in the hollow cubical cavity and let to solidify into a solid. This solid which is largely void free is then used to investigate the melting of solid PCM in the experiments.

Rubitherm's RT35HC [208] is the selected paraffin to validate the numerical model. RT35HC has a melting temperature of 35 °C and stores approximately 240 KJ/kg in the form of sensible and latent heat [208]. The phase change is characterized by a peak inside melting range between the temperatures 34 °C and 36 °C. This organic PCM does not exhibit any supercooling but is inflammable. The material properties documented by Rubitherm in the technical specification of the RT35HC are not sufficient to implement them into a numerical model. Therefore, Rösler [79] measured the physical properties like specific heat capacity, density, thermal diffusivity and dynamic viscosity of RT35HC over a large temperature range. These measured material properties of the paraffin RT35HC are adapted in this work.

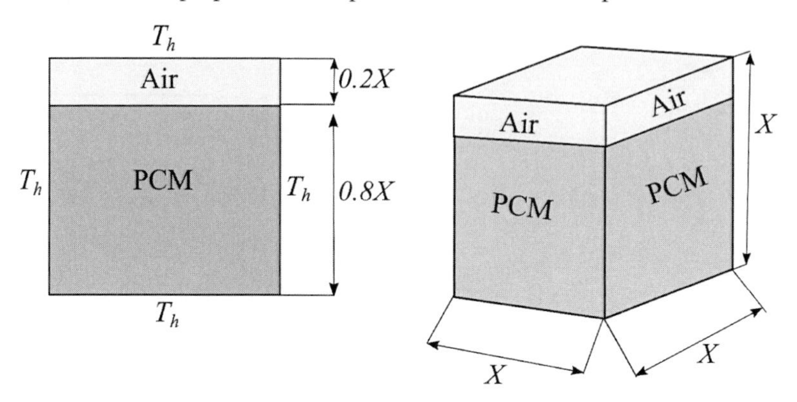

Figure 6.3: Boundary conditions and dimensions of the cubical cavity filled with 80 % PCM and 20 % air.

For experiments, the cubical cavity is filled with 80 % of solid RT35HC and the rest 20 % with air as in Fig. 6.3. During the experiment, the side walls are heated with a temperature T_h = 40.5 °C as depicted in Fig. 6.3. The internal temperature of cubical cavity at the beginning of the experiment is T_{start} = 29.5 °C. Before the experiment, the thermostat supplies the necessary start temperature T_{start} around the cubical cavity. At this point, the complete experimental set-up is left undisturbed until the PCM in the cubical cavity achieves a steady state initial temperature T_{start}. Hereafter, the bypass mode is turned on to preheat the HTF and also to disconnect the cavity from the thermostat. Now, the thermostat is operated to achieve the desired temperature T_h after some time. When the HTF inside the thermostat achieves a steady desired temperature T_h, the flow of HTF around the cavity with a temperature is allowed by switch off the bypass mode. Alongside, a voltage of 12 V and 0.72 A of current is supplied by the power supply unit resulting in 8.64 W of power on the front and back walls of the cavity. The solid PCM inside the cubical cavity gradually melts with time and solid settling takes place. Now, the timely captured pictures depict the melting of solid and increase in the PCM volume inside the cavity as in Fig. 6.4.

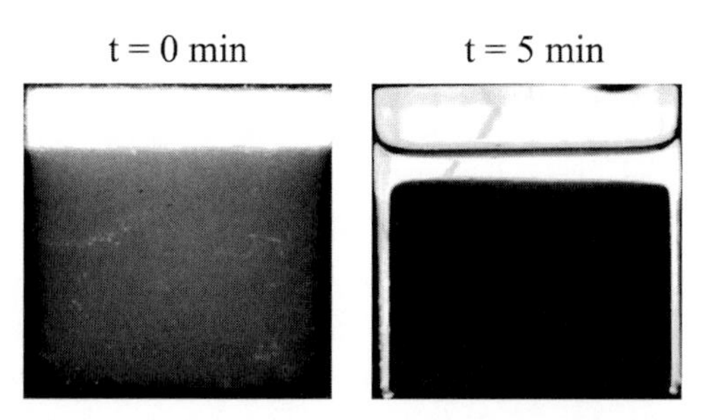

Figure 6.4: Captured photographs of the cubical cavity filled with paraffin RT35HC and air at times t = 0 min and t = 5 min.

The surface tension between the liquid PCM and the air forms a capillary meniscus with the cavity wall. The solid PCM due to its higher density settles at the bottom of the cubical cavity as seen in Fig. 6.4. The nature of solid settling depends on the porosity of solid PCM and the interfacial interactions between PCM and air. The porosity of the solid PCM depends on its preparation, a dense solid with no void spaces is ideal. In the experiments performed, sometimes the solid PCM hanged on the PCM-air interface and settled improperly leading to an asymmetric melting pattern of the solid PCM as depicted in Fig. 6.5.

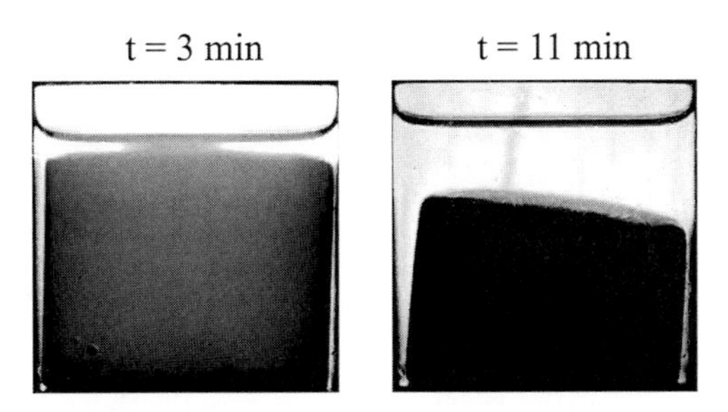

Figure 6.5: The surface interactions between the PCM and air resulting in an improper settling of solid phase shown at times t = 3 min and t = 11 min.

The asymmetric settling of solid PCM in the cubical cavity as seen in Fig. 6.5 can result in an enhanced and an impaired melting rate of solid PCM. This phenomenon cannot be implemented in the numerical model due to its stochastic nature. The range of an error in the melting rates can be calculated by repetitive studies of experiments. The error estimation due to the asymmetric settling of solid PCM lies in the range of 2 % to 10 % of the melting rate, which also depends on the solid skewness during settling.

6.2 Thermometry of phase change in a cylindrical tube

The melting temperature of PCM depends on its constituents and chemical formula. Therefore, different materials have distinctive melting temperatures. As discussed earlier, the optical measurement set-up operates only below 60 °C. Alongside, not all materials having melting temperature lower than 60 °C can be investigated using the described optical measurement set-up as in Fig. 6.2. PCM with opaque liquid phase cannot be investigated using the optical set-up, as the solid-liquid interface cannot be tracked. Thence, a new experimental set-up is adapted to investigate the phase change of PCM with opaque liquid phase and higher melting temperatures. Here, the temperature is measured to understand the natural convection of the liquid PCM during melting.

The measurement set-up developed to analyse the temperature profile of PCM consists of a double wall cylinder, Huber thermostat [209], computer and a resistance thermometer Pt 100. The double wall cylinder permits the flow of HTF between its outer and inner wall enabling the heat transfer between the HTF and the PCM filled in the inner cylinder.

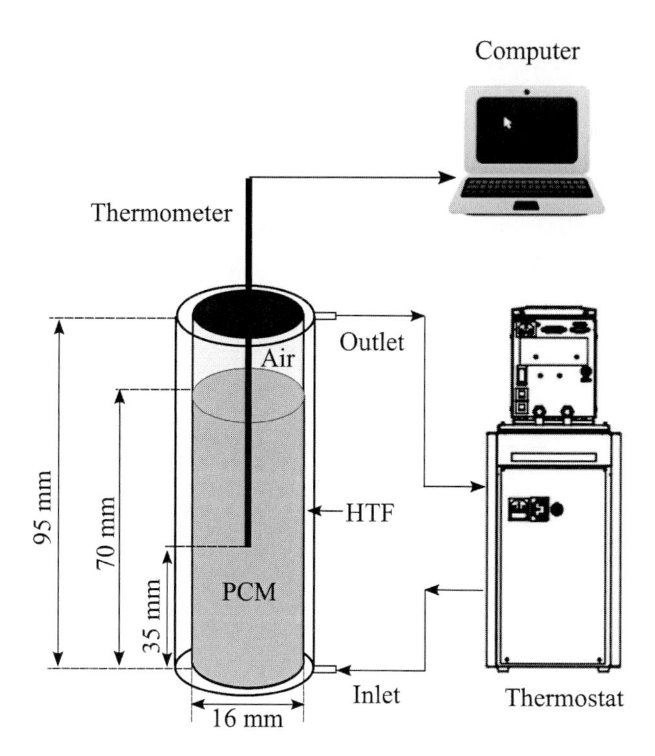

Figure 6.6: Experimental set-up to investigate the temperature profile of the PCM during phase change [209, 210].

Thermometer in the experimental set-up reads the temperature at its point of contact with the material, which is connected to a computer which logs the temperature over time. A high precision Huber ministat 230 thermostat pumps the regulated temperature around the inner cylinder filled with PCM as shown in Fig. 6.6. The cylinder is made of a transparent glass which can bear temperatures above 100 °C. The top wall of cylindrical enclosure is closed using a rubber plug which insulates the PCM from the ambient temperature.

The inner hollow cylinder has a diameter of 16 mm and is 95 mm long. The inner cylinder is filled with 73.68 % of solid PCM, which has a measured height of 70 mm from the cylinder bottom. The PCM used in these experiments is an inorganic salt hydrate, magnesium chloride hexa-hydrate ($MgCl_2 \cdot 6H_2O$) with a melting temperature 115 °C. The measure mass of the PCM filled in the inner cylinder for the experiments is 0.022 kg. The thermometer is placed exactly at the midpoint of the PCM which is at distance of 35 mm from the bottom of the cylinder. The measured temperature at the centroid of PCM is plotted against time. A small air cavity is accommodated between the PCM and rubber plug surface to resist thermal stresses due to volumetric expansion. This air cavity has a height of 20 mm. The material properties of the PCM $MgCl_2 \cdot 6H_2O$ are not measured in the present work. Rather, the materials properties are incorporated from the published in-house measurements performed to measure the thermo-physical properties of $MgCl_2 \cdot 6H_2O$ [211].

In experiments, the cylindrical tube with PCM is initially preheated to a steady-state start temperature T_{start} using thermostat's HTF. Later the hot temperature T_h can be adjusted on the thermostat, which results in a rapid stepwise heating of thermostat's HTF. The heated HTF flows around the cylinder tube filled with PCM. After the hot temperature T_h is adjusted on the thermostat's screen, the internal temperature of the fluid in HTF shoots up to a value greater than T_h. This occurs due to the rapid heating of HTF by the thermostat, which after a while is adjusted to the desired hot temperature T_h. The temperature shoot up in the first few minutes of the experiment is evidential in Fig. 6.8, where the temperature of HTF shoots up to a value of 137 °C despite the desired hot temperature T_h = 130 °C.

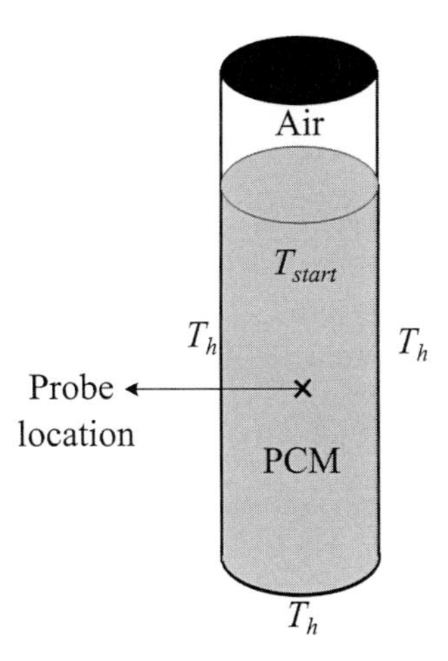

Figure 6.7: Experimental set-up's boundary conditions of the thermometry of phase change of PCM in a cylindrical tube.

The heat flux from the heated cylinder wall results in melting of solid PCM inside the cylindrical tube. Temperature is constantly read from the resistance offered by thermocouple placed at the centre. Accordingly, the temperature of HTF inside the thermostat is also measured and plotted against time. The temperature versus time plot depicts the rise of the PCM internal temperature at probe location as shown in Fig. 6.7. In Fig. 6.8, the difference between the final temperature of the thermostat's HTF and the probe location in PCM are caused by the heat losses. The heat losses could approximately influence the temperature by a magnitude of 1 °C. The start temperature T_{start} in the experiments is 100 °C and the hot surface temperature T_h = 130 °C. It takes more than 30 minutes to achieve a steady state for temperature difference of 30 °C. Therefore, the experiments were performed upto 42 minutes until a full steady-state has been achieved. Different repetitive measurements are performed to conclude the uncertainties in the measurement techniques. Here, it is observed that the reproducibility of these experimental results range within an error of 1 % for melting time.

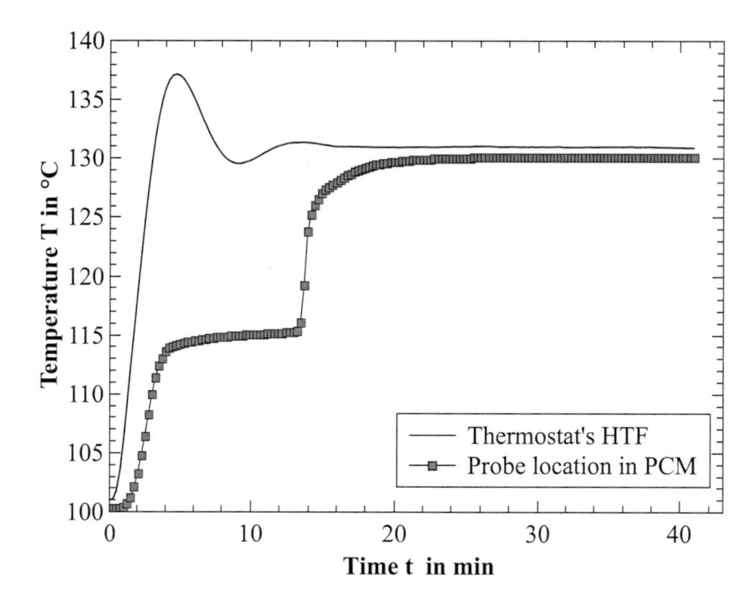

Figure 6.8: The time dependent temperature profile of PCM at the probe location and also the thermostat's internal temperature profile.

The thermometric study of PCM phase change provides a very fine estimate of the liquid PCM convection inside the cylindrical enclosure. In the present work, these experimental results are validated with the numerical model to understand the role of natural convection in a melting process in a cylindrical capsule.

6.3 Investigation of conjugate heat transfer for thermal storage systems

Experimental investigation of the phase change in a cavity or a cylinder just provides an estimate about the heat transfer inside the capsules. In a macro-encapsulated LHTES, the heat transfer between PCM capsule to HTF and HTF to capsule wall also play a major role. Generally, for high temperature macro-encapsulated LHTES units strong opaque materials are used for encapsulation. Therefore, it is not possible to analyse the melting of PCM in a capsule influenced by HTF only with a camera. Therefore, a representative row of capsules in a storage unit for normal temperatures is built to study the heat transfer between the PCM capsules and the HTF. Here, the heat transfer between the HTF to capsule wall and capsule wall to PCM is investigated including the solid-liquid phase change in PCM.

The developed experimental set-up consists of three cylindrical PCM capsules with external diameter 50 mm and length 65 mm. These capsules having a wall thickness of 2 mm are made of borosilicate glass. They are placed in a rectangular cuboid container, which accommodates the flow of the HTF around these cylindrical capsules. The rectangular cuboid container is made of acrylic glass with interior dimensions 0.42 m $\times$ 0.07 m $\times$ 0.07 m. This container is connected to an inlet and outlet through a perforated acrylic glass plate, which enables the homogeneous distribution of the HTF flow.

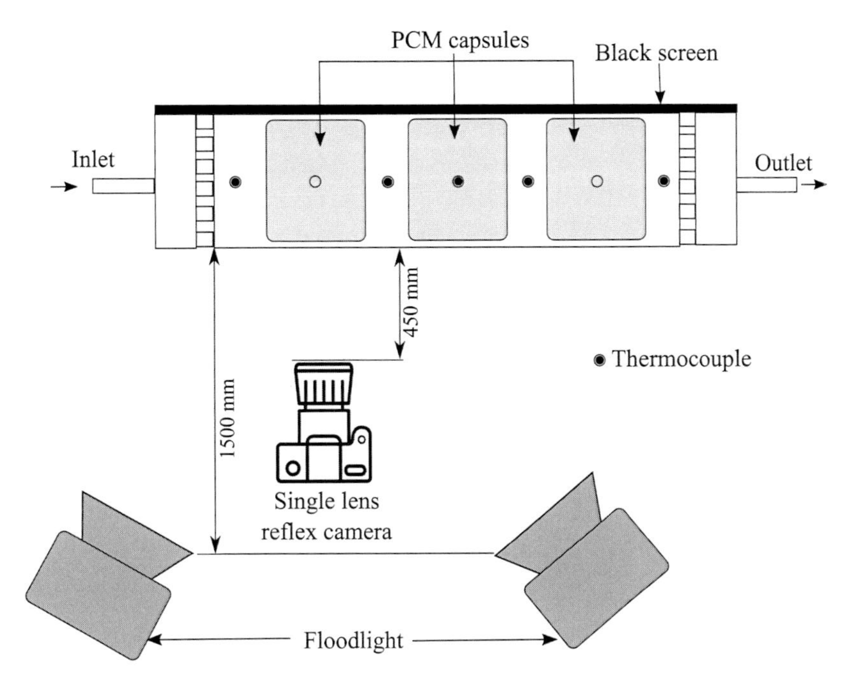

Figure 6.9: Top view of the experimental set-up to investigate the conjugate heat transfer between PCM capsules and the HTF in a thermal storage system.

Thermocouples used to measure the temperature are placed at different positions in HTF and in a PCM capsule to understand the thermal distribution in the experimental set-up. Thermocouples employed to measure the temperature in the flow are placed at different heights to measure the vertical temperature distribution. Because for a very slow flow rate, due to settling of cold HTF at the bottom of the domain resulting in a density layered flow. Thermocouple inside the middle capsule is placed exactly at the centre of the PCM. The inclusion of thermocouple in the solid PCM restricts the settling of solid PCM to the bottom of capsule during melting. Therefore in the middle PCM capsule, close contact melting does not take place. The flow of HTF in the cuboid container through the perforated acrylic glass plates originates from the inlet. The inlet is connected to a Huber ministat 230 thermostat through a buffer container [209]. The outlet is also connected to thermostat, where the HTF from the outlet again heats up before entering the experimental domain through inlet. Pansonic lumix GH4 single-lens reflex camera is placed exactly opposite to the first PCM capsule at a distance of 450 mm from the domain. Therefore, the camera captures the melting of solid PCM in the first PCM capsule. To enhance the illumination inside the container, two floodlights with maximum power of 500 W are placed 1500 mm away from the experimental set-up. Additionally, a blackscreen is placed behind the experimental domain enabling a better capturing of solid-liquid interface. The flow of HTF into the domain is regulated by a flow rate regulator, which is governed by a computer. Temperature measured at different points in the flow and in the PCM capsule are processed by the same computer. The control valves placed at the inlet and the outlet regulate the flow inside the container. Here, the bypass mode of the control valves restricts the preheated HTF interacting with PCM capsules as shown in

Fig. 6.10. An additional buffer container is arranged to store the preheated HTF during the bypass mode. This container is also used to recirculate the HTF with a constant temperature till a steady-state temperature distribution inside the experimental domain is achieved.

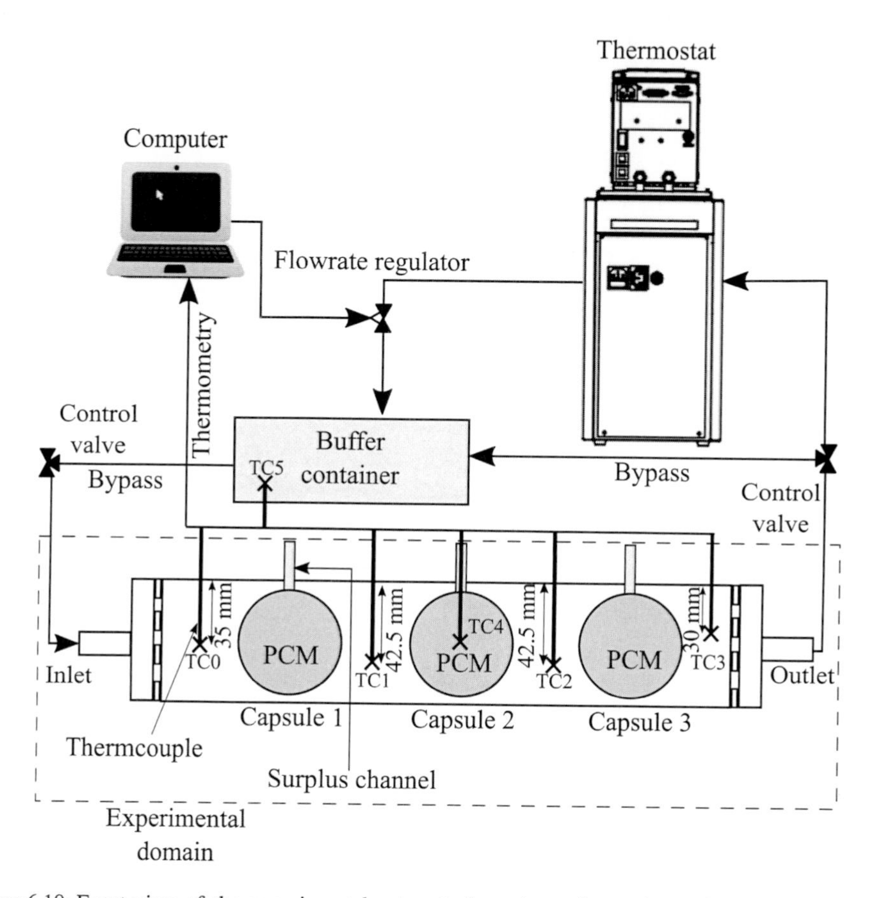

Figure 6.10: Front view of the experimental set-up to investigate the conjugate heat transfer between PCM capsules and the HTF in a thermal storage system [209, 210].

The three cylindrical capsules placed equidistantly from each other in the container are filled with paraffin RT35HC. The cylindrical capsules are built with thin surplus channels to allow the outflow of excess air due to the volumetric expansion of PCM. The cylindrical capsules are filled to 93 % of their volume with solid PCM, which expands to 100 % after melting. These extending channels above the first and last capsules are left open as seen in Fig. 6.10. For experimental investigations, water is employed as HTF due a good balance between affordability and thermal conductivity. Initially, the preheated water at a temperature T_{start} = 29.5 °C is pumped into the experimental domain with a volumetric flow rate $\dot{V}$ = 60 l/h. Here, the paraffin RT35HC in the cylindrical capsules remain solid at 29.5 °C due to its melting temperature at 35 °C. At this stage, the experimental set-up is left unchanged till the whole experimental domain achieve the steady-state temperature T_{start} = 29.5 °C.

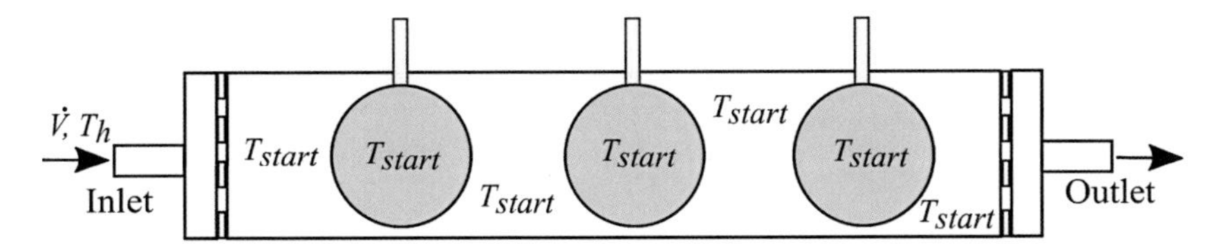

Figure 6.11: Boundary conditions of the experimental set-up to investigate the conjugate heat transfer between PCM capsules and the HTF in a thermal storage system.

Later, the bypass mode of the control valves at the inlet and outlet is turned on alongside adjusting the thermostat temperature to the desired hot temperature T_h = 39.5 °C. During the bypass mode, the inflow of the HTF from thermostat into the experimental domain is restricted. But, the HTF from the thermostat with a temperature T_h = 39.5 °C is pre-stored in the buffer container. The temperature in the buffer container is measured by thermocouple TC5. When the temperature inside the buffer container and thermostat reaches a steady state temperature T_h = 39.5 °C, the bypass mode is turned off to a flow mode. In the flow mode, HTF with a temperature of T_h = 39.5 °C flows with a regulated flow rate of $\dot{V}$ = 60 l/h into the experimental domain. Alongside, the camera placed opposite to the capsule 1 starts capturing the melting of RT35HC. Whereas, the melting of RT35HC in the capsule 2 is measured by recording the temperature at the centre of PCM by thermocouple TC4.

Figure 6.12: Captured photographs depicting the melting of paraffin RT35HC in the cylindrical capsule 1 of experimental set-up at times t = 15 min and t = 50 min.

The photographs are captured at a time interval of t = 2min to evaluate an accurate melting rate of the RT35HC in the capsule 1 as shown in Fig. 6.12. As seen in Fig. 6.13, the temperature at the thermocouple TC4 inside the solid PCM of capsule 2 falls down after t = 75 min. This occurs when the solid PCM till the tip of thermocouple is melted and the unmelted solid PCM above the sensor slips down to the tip of the thermocouple. Later, the temperature rises when the solid PCM detaches from the thermocouple tip. The slippage and detachment of solid PCM from thermocouple took place at different instances in the experiments performed. Due to its chaotic nature, the temperature fall in the highlighted box of Fig. 6.13 is not implemented in the numerical model.

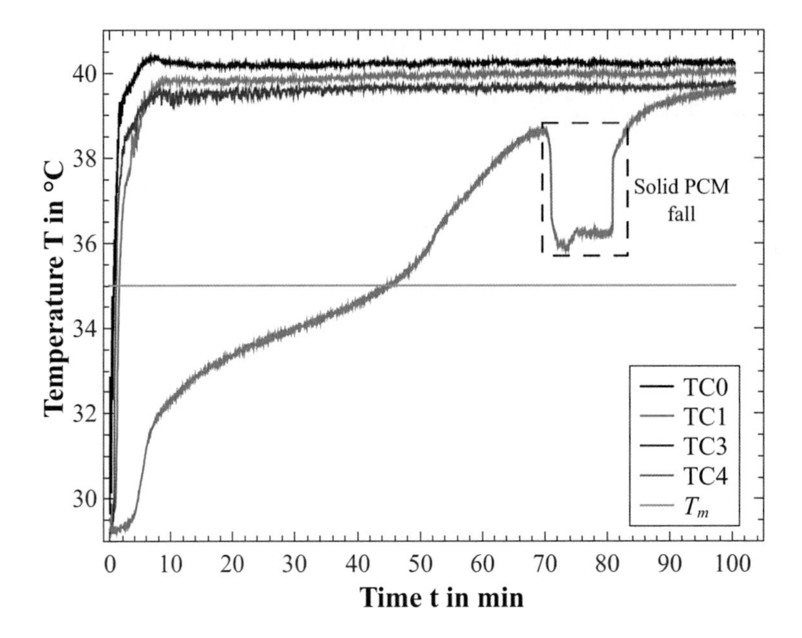

Figure 6.13: Temperature distribution of the experimental domain measured though thermocouples affixed inside the HTF and PCM capsule 2.

6.4 Thermal charging and discharging of laboratory-scale latent heat thermal energy storage

Experimental set-up as shown in Fig. 6.11 to study heat transfer is not substantially representative of a whole LHTES unit. Therefore, a laboratory scale thermal energy storage unit representative of a very large mobile LHTES unit is built. In addition to an efficient thermal charging and discharging, operating at temperatures above 70 °C is also the topic of interest here. To operate at such high temperatures, the laboratory scale LHTES employs a PCM with melting temperature greater than 100 °C.

The laboratory-scale LHTES unit is designed to allocate different PCM geometries like cylinders, spheres and plates [212]. The storage container is made of stainless steel with a usable edge length of 250 mm and height 540 mm. This storage container is connected to inlet and outlet through a perforated aluminium plate as shown in Fig. 6.16. Homogeneous flow of HTF in the storage container enhances the thermal efficiency of storage unit through a well distributed heat flux. The storage container is symmetrical from top to bottom, which allows the flow of HTF in both the directions. Temperature in the storage container is measured by 24 thermocouples at different positions. These thermocouples are placed at regular distances on 4 planar sections of the storage container. Additionally, thermocouples are also placed at the top and bottom flow valves to measure the inflow and outflow temperatures of HTF as shown in Fig. 6.14. The complete storage unit is insulated with an aluminium laminated mat, CLIMACOVER and an insulation wool, ROCKWOOL 800 with a thickness of 100 mm. The storage unit is connected to a thermostat REGLOPLAS 150S to pump the HTF into the container. In experimental set-up, the mass flow rate of HTF is regulated using a control valve, which is connected to a computer. The flow of HTF into the storage unit is governed by control valves V1, V2, V3 and V4.

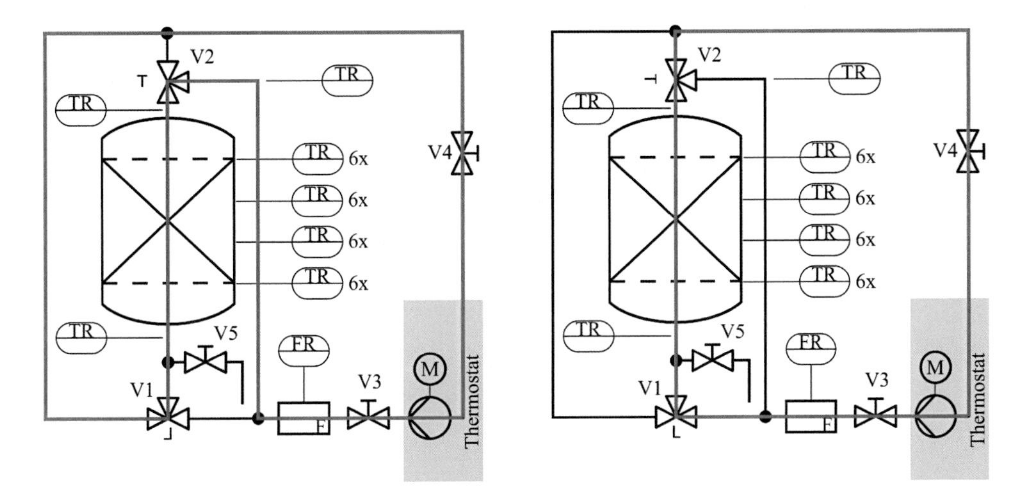

Figure 6.14: Schematic representation of the HTF's flow in the storage unit from top to bottom shown on left side and vice versa on the right side [212].

The storage container is filled with cylindrical capsules filled with salt-hydrate $MgCl_2 \cdot 6H_2O$. The PCM capsules are made of aluminium with an external diameter of 40 mm and a wall thickness of 3 mm. The cylindrical capsules are sealed with fluorinated rubber gasket and the capsule lid is locked with screws as seen in Fig. 6.15. The capsule internal surface is anodized to resist the corrosion effects from the chloride and water in the salt-hydrate PCM.

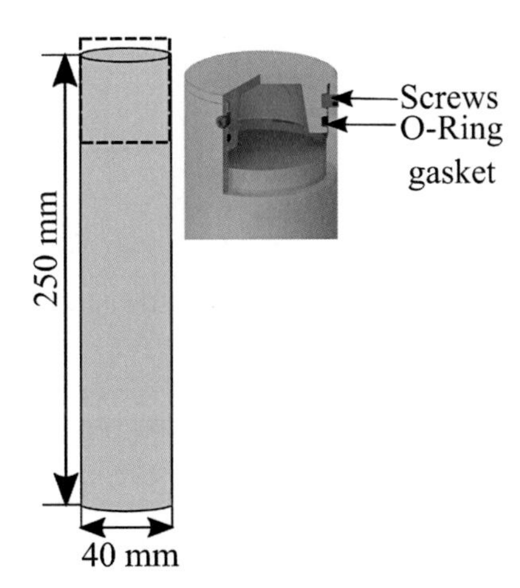

Figure 6.15: Pictorial depiction of cylindrical capsule developed for the present storage unit [212].

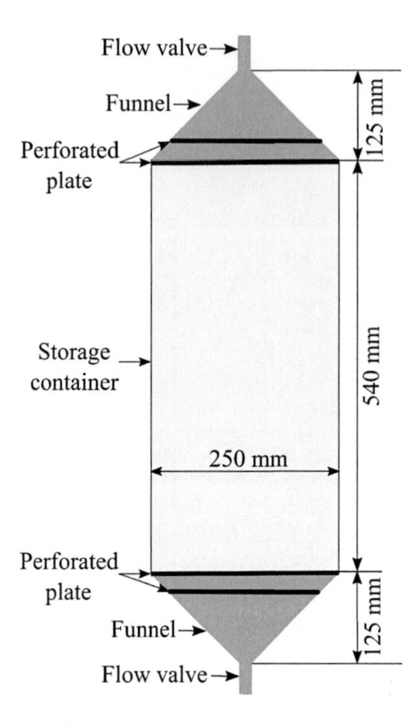

Figure 6.16: 2-D planar depiction of the laboratory scale LHTES developed to operate at temperatures above 100 °C.

The capsules are filled 95 % of its volume with liquid $MgCl_2 \cdot 6H_2O$. The PCM capsules are placed in two sets one above another. Each set of the capsules consists of 36 capsules arranged 6×6 in lateral directions. Thereby, a total of 72 capsules are arranged in two rows with the total packing height of 500 mm. The storage unit is thermally charged and discharged by the flow of HTF (thermal oil) in the storage container. To benefit from the latent heat storage capacity, the charging and discharging of the LHTES unit is operated at a small offset from the melting temperature. Thermal efficiency of the LHTES unit is studied by operating at different operating flow directions. The charging of the LHTES unit by the flow of HTF from the top to bottom and discharging by the flow from bottom to top is found efficient [212]. Otherwise, the charging and discharging process is approximately two times longer than the efficient way. The efficiency is strongly influenced by the convective mixing of the hot and cold HTF in the storage unit.

For charging and discharging, the thermostat follows a ramping procedure to achieve the desired temperature from the start temperature as observed in Fig. 6.17. In Fig. 6.17, a temperature difference ΔT = 30 °C is employed to charge and discharge the laboratory scale LHTES unit at a mass flow rate of 5 kg/min. For both charging and discharging, the LHTES unit achieves a quasi steady state after t = 120 min. The fluctuations in the temperature plot are characterized by the wavy temperature control of the thermostat. Further, different mass flow rates of 2.5 kg/min, 5 kg/min and 10 kg/min and temperature differences ΔT = 20 °C and ΔT = 30 °C are employed to investigate the thermal response of LHTES unit. Accordingly, the thermal power dependent on the mass flow rates, temperature differences and cycle stability is studied. In the present work, the experimental findings are used to validate the LTNE model.

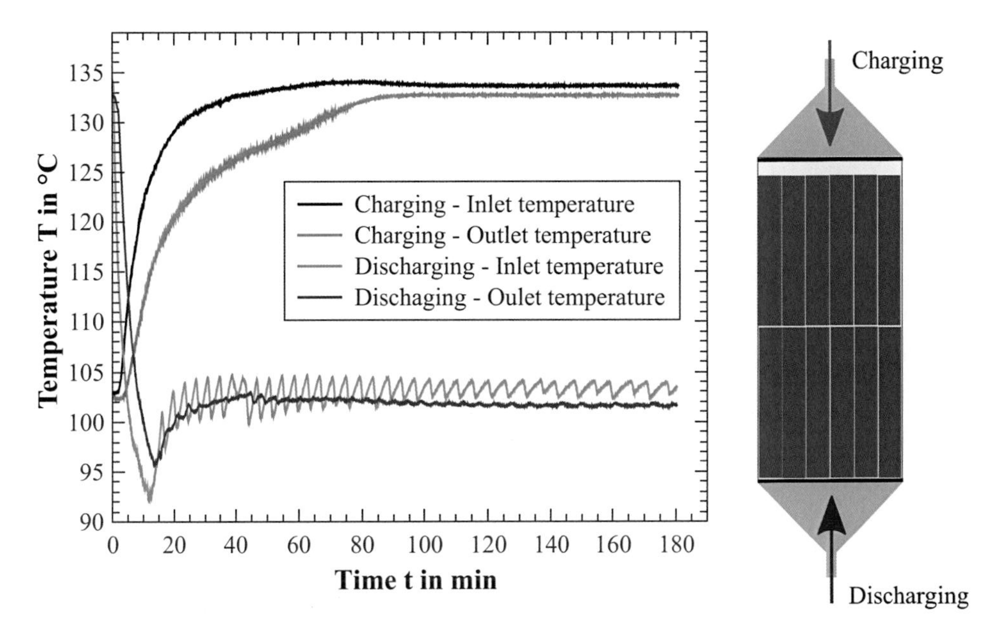

Figure 6.17: On the left, temperature distribution at the inlet and outlet with a mass flow rate of 5 kg/min. The right side picture depicts the efficient flow directions for charging and discharging of the LHTES unit.

7 Validation of numerical models with experimental results

Numerical modelling of physical processes is prolific with the validation of numerical results with experiments. In this work, diversely developed numerical models employed to represent the thermo-fluidic processes in LHTES units are validated with experimental results. To validate the numerical models, numerical and experimental set-up are applied with similar configuration. The numerical model for solid-liquid phase change is initially validated with experimental results. Simulation parameters of the numerical model are optimized for a test case with simple convective solid-liquid phase change.

After validating the simple numerical model with convection, the multiphase numerical model for the solid-liquid phase change in PCM including air in PCM capsule is validated with experimental results. The validated numerical model ensures to estimate the melting of PCM correctly. This complex numerical model due to its high computational effort is not feasible to study sensitivity analysis. Therefore, simplified convection model solving only PCM-phase is developed and validated with experiments. This validation establishes accuracy of the simple convective model with experimental results. However, these models are limited to study the heat transfer in a latent heat thermal storage. The effective thermal conductivity in an enclosed container accounts the magnitude of the convection and close contact melting. The adapted effective thermal conductivity for a specific PCM and geometry type is implemented for a large LHTES unit with PCM capsule-stacks.

To understand the potential of numerical modelling and also to investigate detailed heat transfer, a coupled convective numerical model is developed. The coupled numerical model thereby solves the solid-liquid phase change including the settling of solid in PCM capsules and also the flow of HTF around a PCM capsule. The heat transfer between the PCM capsule and HTF is modelled using a overlapping grid approach by implementing temperature continuity at the grid-interface. This numerical model is validated with experimental results for thermal charging of three cylindrical capsules in a small scale container. The heat transfer including the fluid flow in a capsule row can be understood by validating the numerical model with experiments. However, due to its detailed representation of the physical phenomenon following high computational expense this model is limited to solve very few number of capsules. Therefore, it cannot be extended to estimate the charging and discharging of a complete thermal storage unit. Hence, a simplified LTNE model is developed to compute the heat transfer and fluid flow in a macro-encapsulated LHTES unit with a homogenized approach. Here, PCM capsule packed bed and HTF in a storage unit are homogenized over a REV. While, the complete thermal storage unit container is divided into two identical grids with different material properties. Both identical grids are treated as porous media with their corresponding volumes. Heat transfer between PCM capsule bed and HTF is implemented by computing the total thermal resistance. The pressure loss of the HTF due to the PCM capsule bed is implemented using Darcy-Forchheimer source term in the momentum equation of HTF.

7.1 Numerical discretization and simulation parameters

The developed numerical models are solved for a definite computational domain. To model solid-liquid phase change, depicting the non linear enthalpy-temperature plot with a defined function is the part of initial assumptions. This is followed by modelling the velocity suppression in PCM solid phase. In the present work, velocity in the solid phase is suppressed using VVM. By employing VVM, a large viscosity value is implemented in PCM solid phase resulting in the fixing of solid. Different viscosity value assumptions for solid phase are adapted and tested to achieve a consistent phenomenon. The viscosity of the solid phase is implemented in Eq. 5.26 and Eq. 5.58 with a choice of A value. The adapted value for A implements a temperature dependent viscosity relation for solid-liquid phase change. Accordingly, the solid phase in the computational domain takes a very large viscosity value and behaves like a high viscous material.

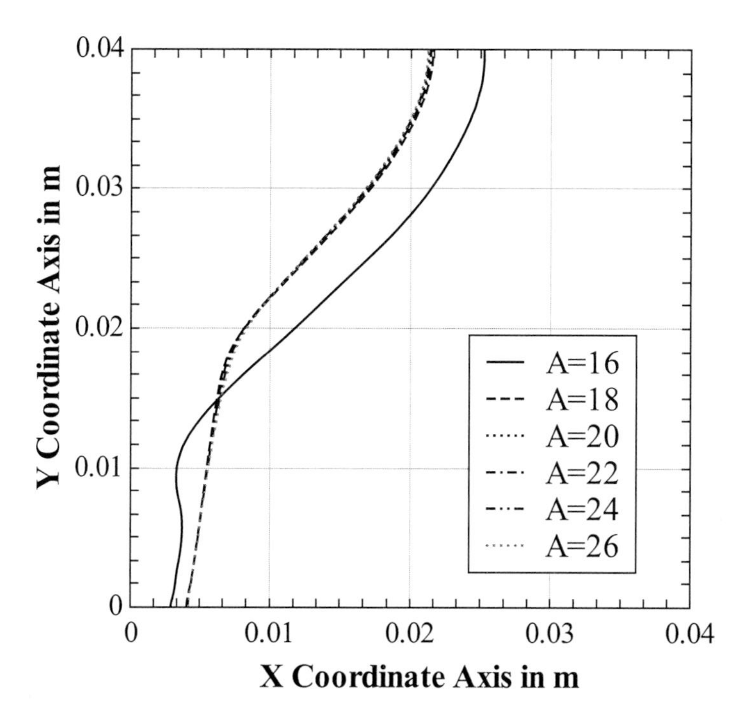

Figure 7.1: Comparison of melt fronts of paraffin RT35HC with different consideration of solid viscosities for a heated left wall case.

The choice of different viscous exponential coefficients A yield different viscosity values in solid phase. By adapting different viscous exponential coefficients between A = 16 and 26, the solid viscosity value of paraffin RT35HC ranges between 3.9×10^4 and 8.6×10^8 Pa·s. Numerical model for solid-liquid phase change employing the VVM is solved for cubical cavity test case [79]. Here, for a cubical cavity with dimensions $40\times40\times40$ mm^3 filled with paraffin RT35HC, the left wall is heated with 42 °C. The internal and right wall temperature remain are at 30 °C with an adiabatic condition for top and bottom walls. As seen in Fig. 7.1, the change in the convective melt front after the value of $A = 18$ is negligible. Here, the amount of resistance offered by the solid phase after this value has no influence on the melting pattern of the paraffin. Therefore, in this work a value of $A = 18$ is employed for all further simulations with paraffin RT35HC.

Different discretization methods are studied for a specific problem definition. The spatial discretization of the domain is an important step to approach any physical solution. Here, different grid configurations for the cubical cavity geometry are investigated. Due to a quasi 2D computational domain, the discretization domain consists of only one finite volume in Z direction. Therefore, different grid configurations in X and Y directions with 100×100 cells, 150×150 cells and 200×200 cells are investigated to evaluate grid dependency of numerical solution.

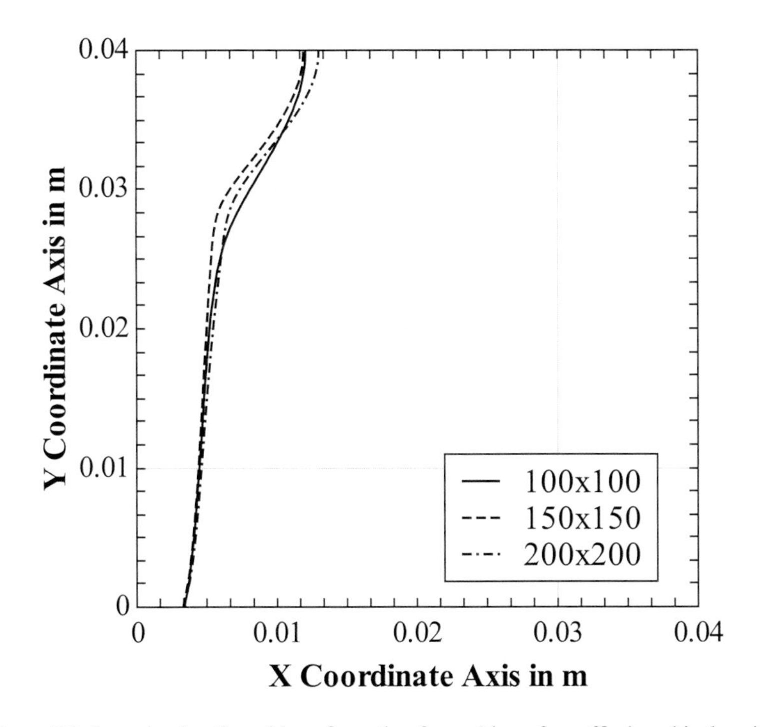

Figure 7.2: Investigating the grid configuration for melting of paraffin in cubical cavity.

A cell expansion approach is employed, where the cells on the boundary layer of the geometry are fine discretized. The boundary layer cells have a very low cell volume, where an expansion coefficient is defined resulting in a gradual increase of cell volume till centre. Solutions from different grid configurations are compared and depicted in Fig. 7.2.

The melt fronts obtained from different grid configurations are shown in Fig. 7.2. Here, the nature of melt fronts look comparable between different computational grids. However for the cell configuration with 100×100 cells, melt front looks linear than from other grid configurations. Therefore, the solution from grid configuration 150×150 cells is efficient, as it delivers an optimal accuracy with a decent computational effort. Therefore in the present work, for a cubical cavity test case a computational grid with 150×150 cells is employed.

7.1.1 Validating numerical method for a convection driven melting problem

Initially, the numerical model developed to solve solid-liquid phase change is validated with experimental results from Rösler [79]. In the present work to validate initial model parameters no additional experiments were executed. The numerical set-up with 150×150 cells and a solid viscosity coefficient A = 18 is developed to computed solid-liquid phase change of paraffin RT35HC in a cubical cavity with dimensions 40×40×40 mm^3.

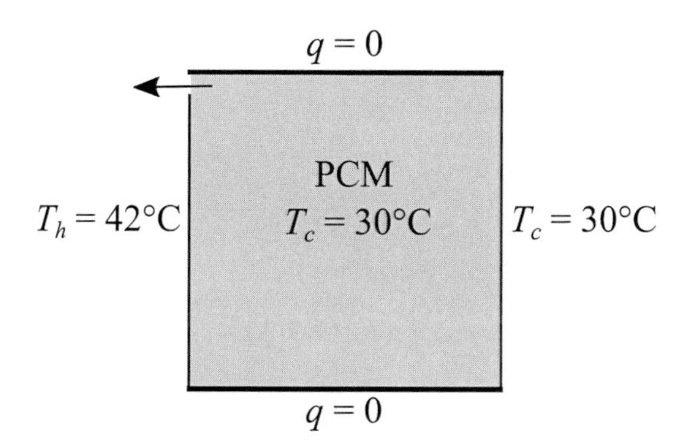

Figure 7.3: Depiction of numerical set-up replicating the experiments employed to validate the numerical model for solid-liquid phase change [79].

In a recent work, the material properties of paraffin RT35HC were also measured along with experiments [79]. Therefore, it is refrained from measuring the material properties of paraffin RT35HC here. Further details on the material properties are documented, see [79]. The cubical cavity is completely filled with paraffin and an outlet is allocated to allow the outflow of liquid PCM from density change. The left wall of cubical cavity is heated with T_h = 42 °C and the right wall is heated with T_c = 30 °C. At the beginning of experiment, internal temperature of cubical cavity is maintained equal to the right wall temperature. Here, the solid-liquid interface from numerical simulations is compared with experiments [79]. Numerical results for melting of paraffin RT35HC has displayed a very good acceptance with experimental results. Here, the solid-liquid phase change of PCM RT35HC driven from natural convection is computed. The nature of convection including the suppression of velocity in solid phase is very well represented by the numerical model. An iterative optimum approach employed to represent the non linear enthalpy-temperature relationship has shown a very good presentation of reality.

As seen in Fig. 7.4, the simulation results show an excellent acceptance with the experimental results at t = 1 h and 2 h. The deviation between the phase fronts at t = 3 h and 4 h is negligible due to an additional heat flux from top and bottom despite adiabatic assumption. In experiments, a very small heat flux flows into cavity from top and bottom due to the good thermal conductivity of acrylic glass. This drives an additional melting of PCM seen at t = 3 h and 4 h in experiments. In Fig. 7.4, the solid-liquid interface is evaluated over a very thin mushy zone, which is treated as an interface area between the phases. But in the simulation results, an exact midpoint of solid and liquid phases is treated as the solid-liquid interface. Thereby the experimental results in Fig. 7.4 are spread over a thin interface area. In experiments, the mass out flow rate of liquid PCM leaving cubical cavity is not exactly estimated. Therefore this cannot be implemented in the numerical model in an exact manner.

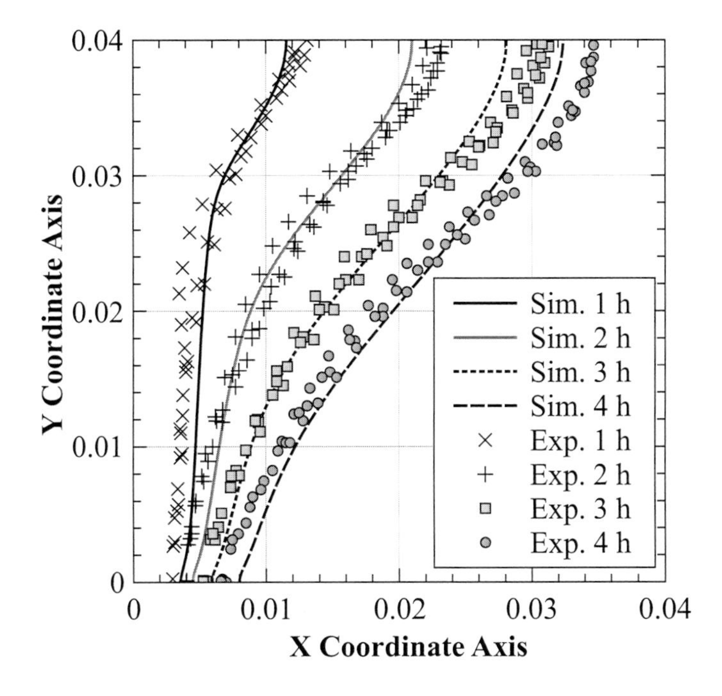

Figure 7.4: Validating the numerical model and parameters for solid-liquid phase change in a cubical cavity of size 0.04 m with experiments [79].

7.1.2 Validating numerical methods for solid settling in PCM capsule

During the charging of a macro-encapsulated LHTES unit, solid PCM inside the PCM capsule settles at the capsule bottom wall. This accelerates PCM melting due to reduced thermal resistance from the capsule bottom. In the present work, two different numerical methods are employed to account solid settling in the numerical model. Implicit method described in the present work is an implementation of solid settling with different density values in each phase. Here, the density difference between the solid and liquid phase of PCM is implemented in the momentum conservation equation with Bousinessq approximation. The large value of viscosity implemented in PCM solid phase resists the movement of solid domain. Due to large viscosity difference between solid and liquid phases, the pressure-velocity convergence of multiphase solid-liquid mixture needs large iteration count [83]. The convergence is adapted using a very small time step obtained from minimal Courant number implemented.

In this work, an explicit approach for solid settling is implemented from force equilibrium method proposed by Asako et al. [65]. Instead of an iterative Newton-Rapshon approach, here a single iteration method is implemented. Where, the solid velocity is computed from the force difference between previous and present time step. The large computational effort from implicit method can be reduced with explicit force equilibrium method. But, due to its explicit treatment the method is limited for only few applications.

In the present work, both methods are employed for cases depending on their operating conditions. Therefore, implicit and explicit methods are compared to illustrate the difference between these models. Accordingly, these numerical models are employed to compute the melting and settling of paraffin RT35HC in a cubical cavity. The numerical results from both the approaches are compared with experimental results [79].

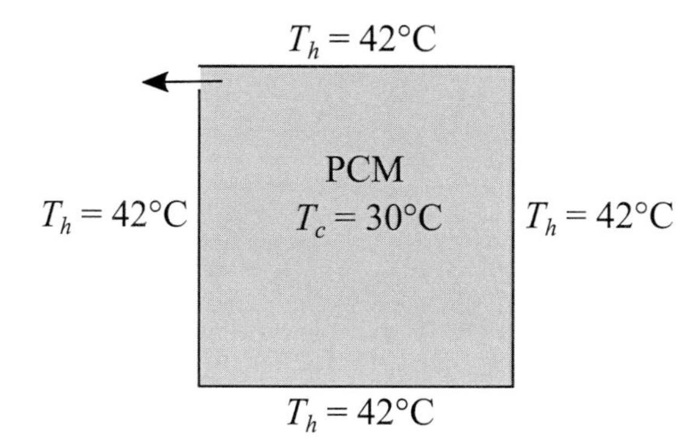

Figure 7.5: Numerical set-up employed to validate the numerical model for solid-liquid phase change including solid settling with experiments [79].

The cubical cavity side walls are heated with a hot wall temperature T_h = 42 °C. At the beginning of experiments, the internal temperature of PCM is adjusted to T_c = 30 °C. An outlet as shown in Fig. 7.5 enables the outflow of excess liquid PCM in the cubical cavity. The cubical cavity is filled completely with solid paraffin RT35HC, which experiences a volumetric change due to melting. Therefore, the outlet in cubical cavity compensates the volumetric change by letting out the excess liquid PCM. The qualitative validation of different settling models for paraffin melting and settling in a cubical cavity with experiments is shown in Fig. 7.6. The explicit force equilibrium method and implicit low time step method have shown very good acceptance with experimental results. The explicit force equilibrium over predicts the melting of solid PCM as seen in Fig. 7.7. Whereas, the implicit approach with a low time step predicts the melting of solid PCM sparsely. Nevertheless, both settling algorithms show a good acceptance with the experimental results.

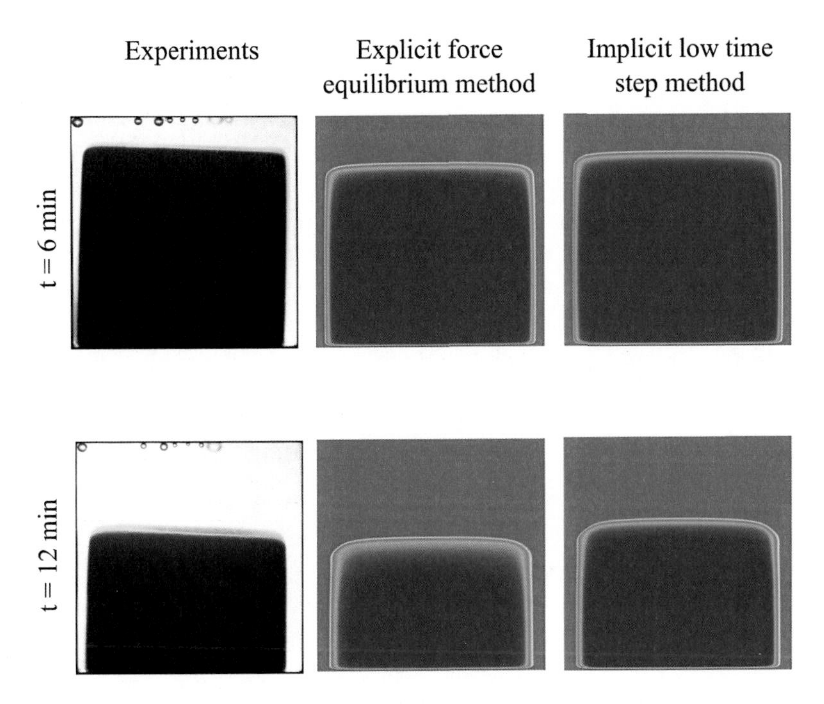

Figure 7.6: Qualitative validation of different settling models with experiments [79].

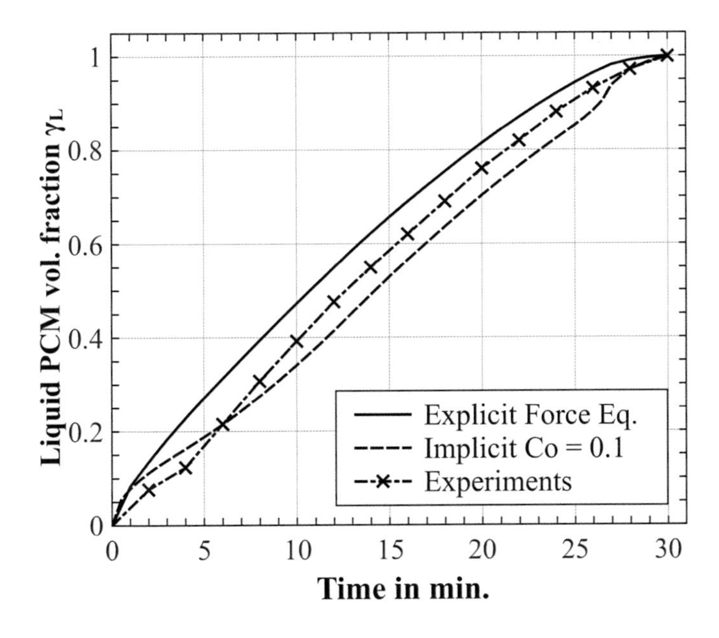

Figure 7.7: Quantitative validation of different numerical models for solid settling with experiments [79].

7.2 Multiphase numerical model for melting of encapsulated PCM

Detailed numerical modelling of solid-liquid phase change in a PCM capsule improves the understanding of heat transfer inside the capsule. The developed multiphase numerical model depicts three phase liquid-solid-gas system in a PCM capsule. The results from numerical simulations are validated with optical measurements for the melting of PCM in a cubical cavity. The problem definition numerically replicates the experimental set-up filled with PCM and air. The heat flow through the copper plates into the cubical cavity is implemented with a temperature boundary condition. In experiments, the cubical cavity is filled with 80 % of PCM RT35HC and the rest with air. Therefore, a similar set-up is employed to compute the solid-liquid phase change in cubical cavity as seen in Fig. 7.8.

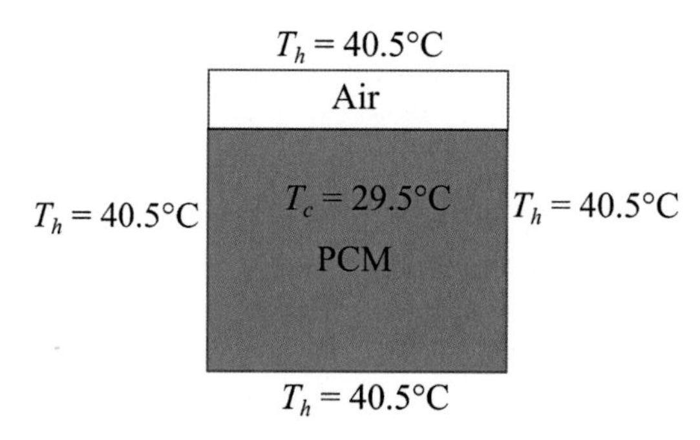

Figure 7.8: Numerical set-up of cubical cavity filled with paraffin RT35HC and air.

A dissimilarity between simulation and experimental set-up is the boundary conditions on the front and back sides of the cubical cavity. For numerical set-up, the front and back sides of cubical cavity are considered symmetry. Whereas in the experiments, a small heat flux flows into the cubical cavity through front and back sides. This minute heat flux from the front and back sides of cubical cavity counters the fixing of solid PCM.

Table 7.1: Material properties of the PCM RT35HC.

Material properties	units	Solid	Liquid
Density (ρ)	kg/m^3	830.9	778.2
Viscosity (μ)	Pa $\cdot$ s	2.8×10^5	0.0044
Specific heat capacity (c)	J/kg.K	5000	2100
Latent heat (L)	J/kg	0	220000
Thermal conductivity (κ)	W/m·K	0.65	0.166

Material properties in Tab. 7.1 are adapted in this work based on the experimental findings from Rösler [79]. The properties implemented in numerical model as in Tab. 7.1 are calculated from an average of independent values over temperature. Additionally, the measured surface tension coefficient between liquid PCM and air phase is 0.026 N/m. The surface tension between liquid and air phase in cavity forms a meniscus with cavity wall. In the numerical model, the capillary effect due to immiscible fluid phases is accounted in the form of wetting. The wetting of the liquid PCM on the cavity wall is achieved by implementing a contact angle.

Due to high wettability of liquid paraffin, an acute contact angle is observed. Therefore in the numerical simulations, a contact angle of 25 ° is implemented between liquid PCM and cavity wall.

In accordance with the numerical set-up in Fig. 7.8 and material properties in Tab. 7.1, the multiphase numerical model for solid-liquid phase change predicts the melting of paraffin RT35HC in a cubical cavity. The results from the numerical simulations are compared with experiments, where the dimensions of solid PCM, melting rate and PCM melting pattern are compared as in Fig. 7.9.

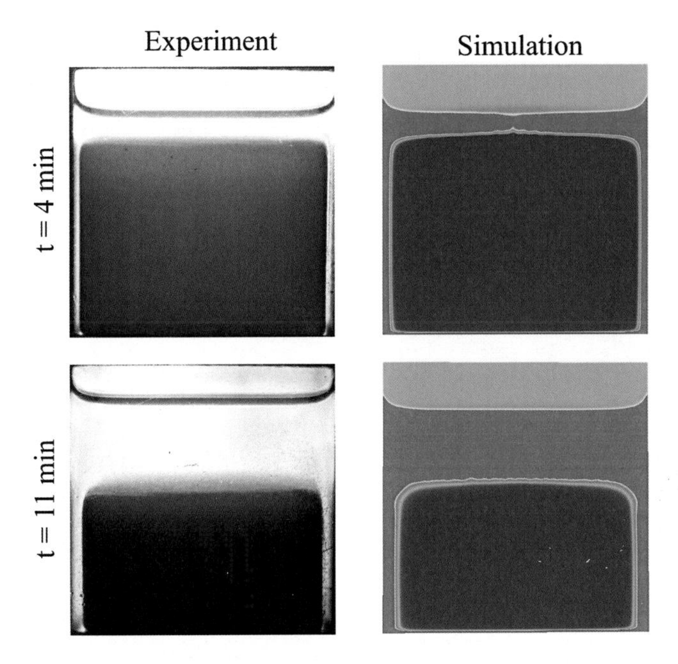

Figure 7.9: Qualitative validation of multiphase numerical model for solid-liquid phase change with experiments.

Results from the numerical simulations have shown a very good acceptance with experiments. The dimensions of solid PCM at time t = 4 min and t = 11 min look very similar to experiments. However, at time t = 11 min the volume fraction of PCM phase inside the cavity looks slightly deviating from the experiments. This deviation originates from a small pressure vent installed in the experiments to relieve the high pressures in cavity. Therefore, in experiments the air outflow through the pressure vent prevents the rupture of front and back plastic discs. To account the realistic pressure change inside the cavity, this outflow of air is not implemented in the numerical simulations. Hereby a small deviation in the volume of air between experiments and simulations can be quantified. The qualitative validation of the PCM melting in a cubical cavity under the presence of air phase is presented in Fig. 7.9. At different time instances the dimensions of solid PCM from the numerical simulations are compared with experiments in Fig. 7.10.

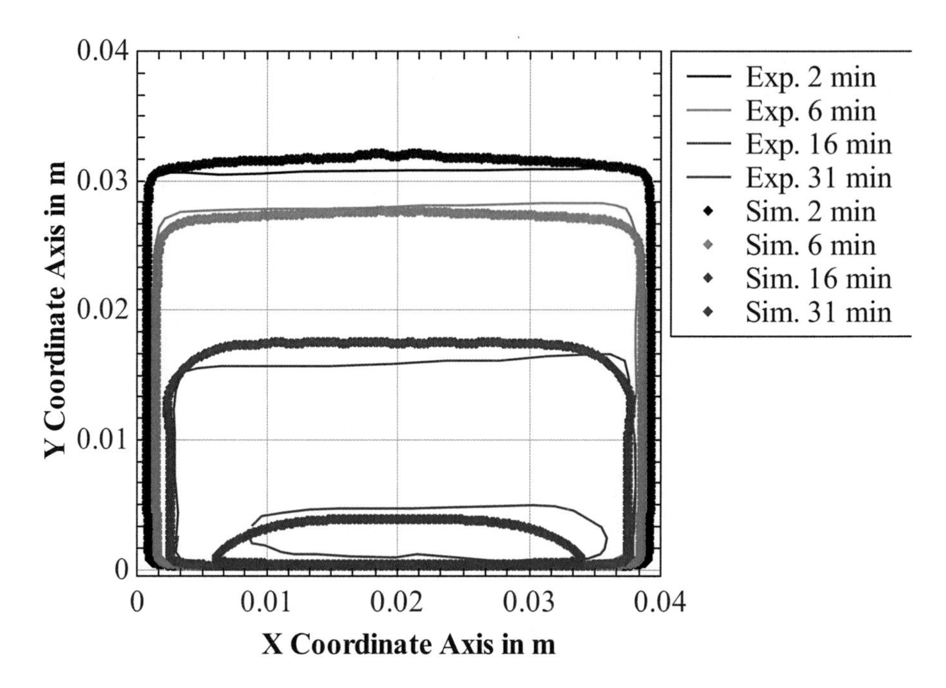

Figure 7.10: Comparison of solid PCM dimensions from experiments and simulations at different times.

As seen in Fig. 7.10, the multiphase numerical model depicts the melting and settling of solid paraffin in cubical cavity very well. Very small deviation at times t = 16 min and t = 31 min can be observed, where the solid PCM melting decelerates in numerical simulations compared to experiments. Such small differences may arise from different assumptions in the numerical modelling and simulations. Some stochastic effects in the experiments cannot be specified in the numerical modelling. For example, in numerical model, a quasi-steady settling velocity is achieved after the solid PCM settles at the bottom of cubical cavity. Also in the numerical model, the movement of the solid PCM in the lateral direction is not implemented. The solid PCM in experiments experience a lateral shift due to the presence of air bubbles in solid PCM. Otherwise, the dimensions of solid PCM in the numerical simulation are very identical to those in experiment.

In addition to qualitative validation, quantitative volumetric melting rate of solid PCM in the cavity is also the topic of interest. For this, the unsteady volumetric melting from numerical simulations is compared with experiments. The volume fraction of liquid PCM inside a cubical cavity is computed from the temperature distribution.

Fig. 7.11 illustrates a very good acceptance for an unsteady comparison of liquid PCM volume fraction γ_L from numerical simulations with experiments. A complete phase change of solid PCM occurs at t = 36 min in experiments and at t = 35 min in simulations. The sharp fluctuations in melting curve of experiment as seen in Fig. 7.11 is driven by the stochastic effects like the presence of bubbles, lateral displacement of solid PCM etc., which are neglected in the numerical model. Until t = 2 min, the melting curve of PCM of numerical simulation overlap with experimental melting curve. A minute deviation of the melting curve from experiments is observed after t = 3 min.

The reproducibility of experimental results depends on the solid PCM settling pattern as discussed in Fig. 6.4 and Fig. 6.5. During the melting of solid PCM, volume fraction of PCM in the capsule increases due to volumetric expansion of liquid PCM phase. The liquid PCM volume fraction in Fig. 7.11 is calculated from fraction of PCM phase inside the cubical cavity. Therefore, the melting rate of PCM also depends on the volume occupied by PCM phase inside the cavity. The deviations in the melting rate of solid PCM can be further studied by comparing the PCM phase from numerical simulations with experiments.

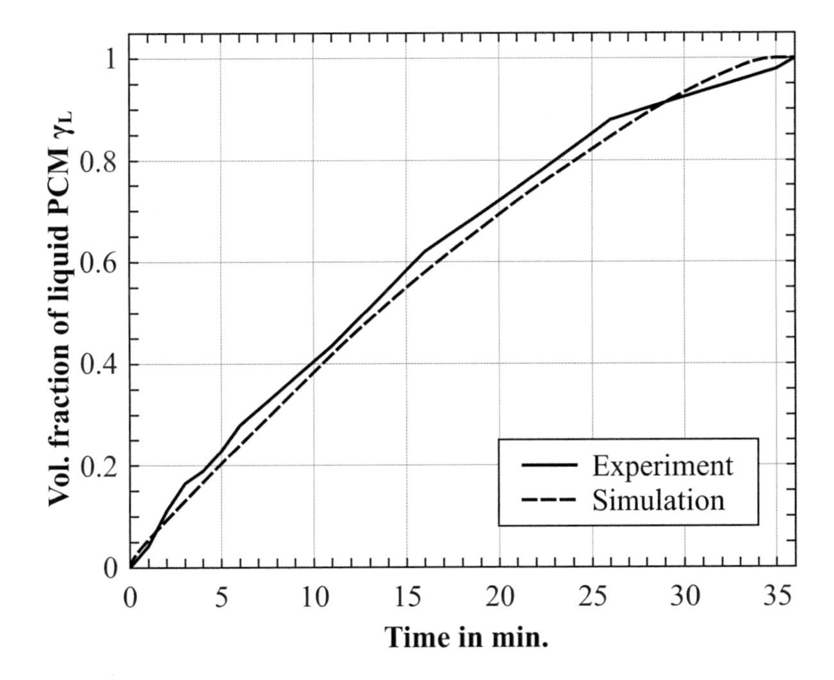

Figure 7.11: Comparison between the volume fraction of liquid PCM from simulation and experiment over time.

In the numerical model, interfacial interactions between liquid PCM and air phase are implemented using surface tension coefficient and contact angle. Compression of air caused from to the expansion of PCM phase pushes the interface in upward direction. In experiments, the movement of PCM-Air interface is influenced by the outflow of excess air as seen in Fig. 7.12. The pressure vent installed in the experimental set-up relieves additional air during the PCM volumetric expansion. Therefore, the minimal deviation between PCM-Air interface from experiments and simulations is influenced by the outflow of the excess air. Nevertheless, capillary meniscus between PCM and air phase is well depicted by the numerical model as seen in Fig. 7.12.

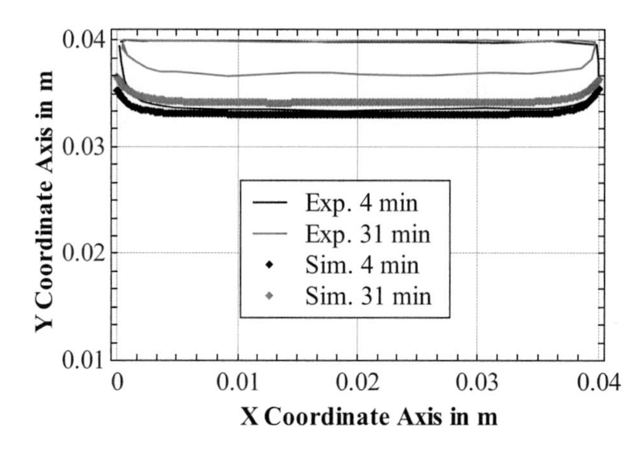

Figure 7.12: Comparison of PCM-Air interface from experiments and numerical simulations at time t = 4 min and t = 31 min of melting process.

In simulations, pressure is built inside the cubical cavity from the volumetric expansion of liquid PCM. The unsteady increase of pressure and PCM volume fraction in the numerical simulations is depicted in Fig. 7.13.

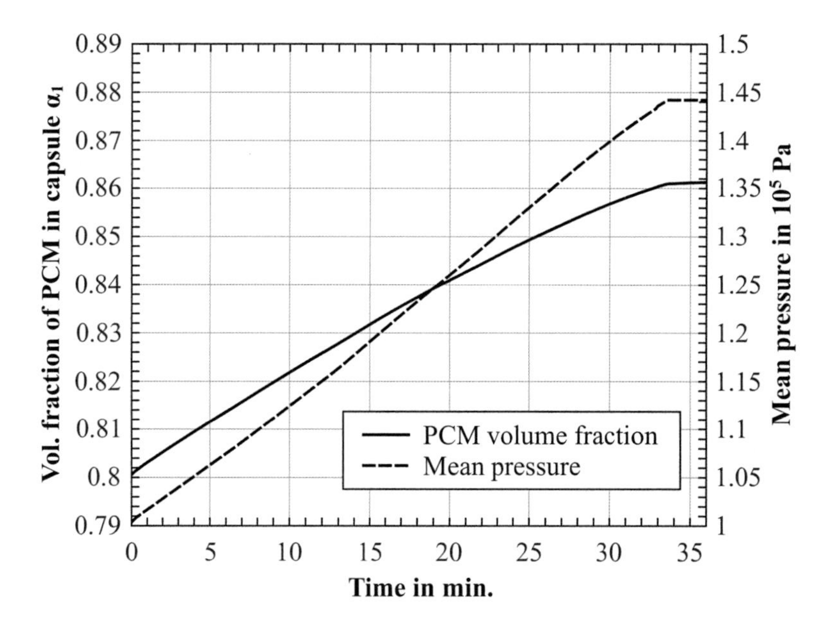

Figure 7.13: Transient increase of PCM volume fraction and mean pressure inside the cubical cavity.

The volumetric increase of PCM phase follows a linear increment over the time. During melting, due to an enclosed cavity the mean pressure inside the cavity rises. After t = 34 min the mean pressure and PCM volume fraction takes a constant value when the melting is complete. Melting and settling of solid PCM in an enclosure can be studied by observing the heat flow from cavity walls. Natural convection of liquid PCM and solid PCM settling play a major role in the melting of solid PCM. Therefore, the heat flow from cavity boundaries is depicted over time in Fig. 7.14.

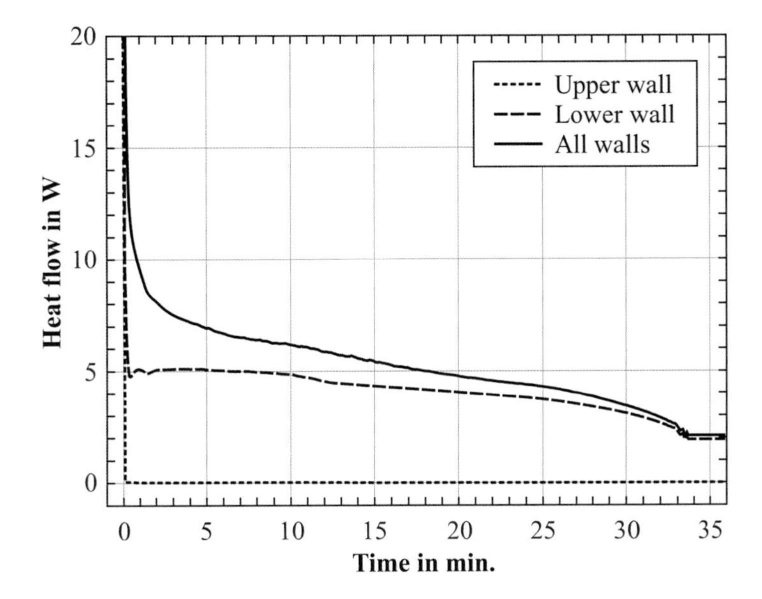

Figure 7.14: Heat flow into the cubical cavity from different walls depicted over time.

As seen in Fig. 7.14, initially from a large temperature gradient, heat flow into the cubical cavity at wall is observed. Later, the heat flow depends on thermal resistances at the wall boundary. Due to air phase, a small portion of heat flows from the upper wall of the cavity i.e. 0.18 % of the total heat flow from all walls. The poor thermal diffusivity of the air builds a large thermal resistance for heat transfer. The major heat flow into cubical cavity takes place from lower wall of the cavity. Due to close contact melting, heat flow from lower wall increases from 37 % of the total heat flow from all walls at t = 0.1 min to 91.67 % by time t = 35.75 min. Here, close contact melting influences the melting of solid PCM to a large extent. Therefore to predict the melting of PCM, it is important to model the settling of solid PCM exactly.

The multiphase numerical model predicts the melting of solid RT35HC in a cubical cavity with air very well. But, complex surface interactions between different phases demand a low operating time step. Therefore, numerical simulations require 40 days to compute the melting of PCM here. To address this, simplified approaches are developed for a low computational time.

7.3 Simplified numerical models for melting of encapsulated PCM

Simulation of solid-liquid phase change with multiphase numerical model is computationally arduous, as it takes the wetting of the liquid PCM, flux segregation, multiphase energy equation and settling of solid PCM into account. Parametric study and variational computations are limited with such a numerical model. Therefore, simplified numerical models are developed to compute the melting of PCM in a capsule with few assumptions. In simplified models, the multiphase system in a PCM capsule is treated as a simplified single PCM phase. The effects of secondary air phase on PCM melting are treated using a convective heat transfer boundary condition. The heat transfer coefficient between PCM and air phase is computed to estimate the heat flow. To compute the heat transfer through PCM-air interface, an independent air phase enclosed in a rectangular cavity with dimensions $8\times40\times40$ mm^3 is employed. At the PCM-air interface the heat transfer boundary condition with the neighbouring PCM solid phase is implemented as seen in Fig. 7.15.

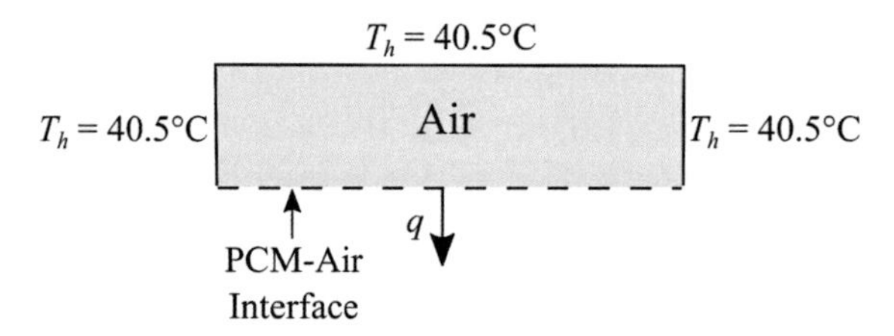

Figure 7.15: Numerical set-up to compute the heat transfer between the PCM and air phase.

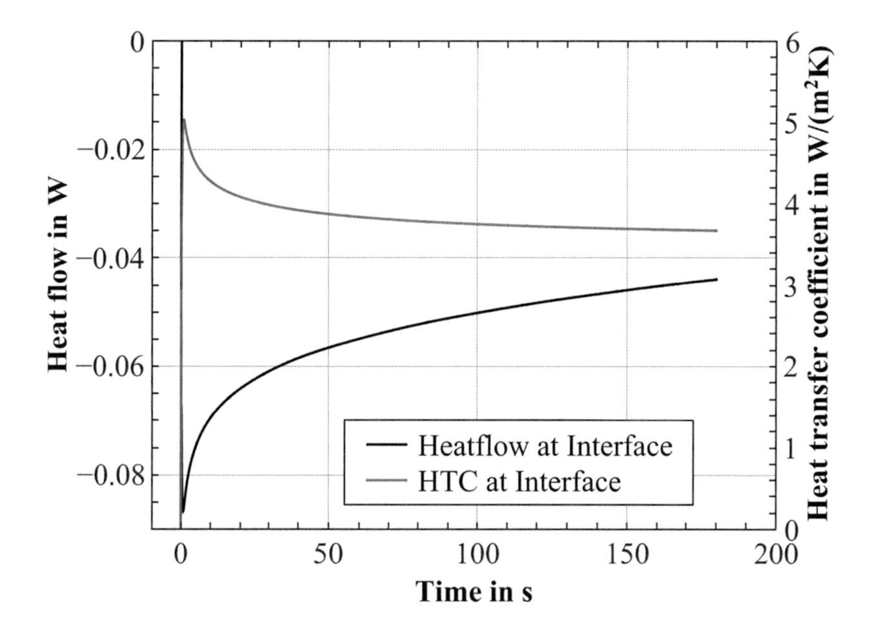

Figure 7.16: Heat flux and heat transfer coefficient between air and PCM phase in a cubical cavity.

The heat flow through PCM-air interface on to the PCM phase is computed. The average heat transfer coefficient is estimated from computed heat flow and temperature difference shown in Fig. 7.16. Due to large temperature difference, a large heat flow is observed at the interface initially. The heat flow then further takes a consistent increment as seen in Fig. 7.16. The corresponding heat transfer coefficient is calculated from temperature difference between the phases and interface area. Initially, the heat transfer coefficient takes a very large value due a large heat flux, which later takes fluctuating values ranging between 3.6 and 5 W/m^{2}K. Accordingly, the average heat transfer coefficient between air and PCM phase in cubical cavity is 3.85 W/m^{2}K. This value corresponds to the heat transfer influenced from the natural convection of air phase. In this work, the simplified models represent the solid-liquid PCM mixture using conservation equations for incompressible flow. Therefore, unlike the multiphase numerical model a constant density for both solid and liquid PCM phases is implemented in the energy equation. Thereby, density change between solid and liquid phase is adjusted in the governing equations to account the correct energy state. This is mathematically represented as [213]

$$\phi_S = \frac{\rho_S}{\rho}\phi \tag{7.1}$$

where, ϕ is any thermo-physical quantity like specific capacity, thermal conductivity and latent heat n the energy equation, ρ_S is the density of solid PCM and ϕ_S is the corresponding thermo-physical quantity of solid PCM phase. For both phases, density ρ takes a constant value in Eq. 5.60.

7.3.1 Simplified convective model

In the present work, simplified numerical models take only PCM phase into account to model the phase change in a PCM capsule. An additional air phase is modelled as a convective boundary condition.

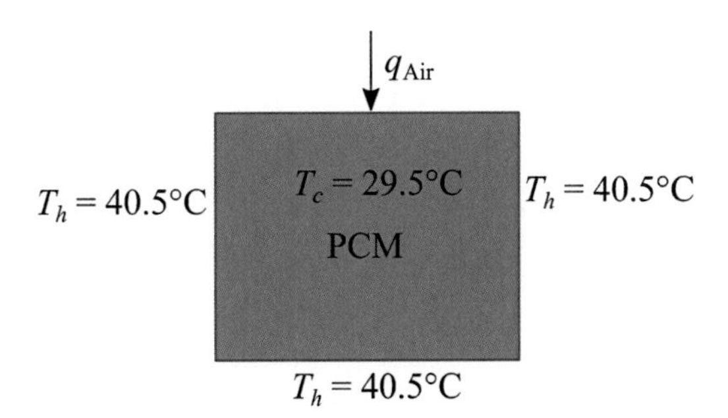

Figure 7.17: Numerical set-up to compute melting of PCM in a cubical cavity with simple convective model.

Average heat transfer coefficient between air and PCM phases is implemented as a boundary condition on the interface. Schematic depiction of numerical set-up to compute the melting of PCM in a cubical cavity is shown in Fig. 7.17. Heat transfer condition with HTF is implemented for the bottom, left and right walls of cubical cavity. Here, for upper wall of PCM domain, a convective heat transfer boundary condition for heat transfer between air and PCM is implemented as

$$q_{Air} = h_{int} \cdot A \cdot \Delta T. \tag{7.2}$$

In Eq. 7.2, q_{Air} is the heat input from air phase onto PCM, h_{int} is the corresponding heat transfer coefficient, A is the surface area between PCM and air phase and ΔT is the temperature difference. The average heat transfer coefficient value at the interface h_{int} = 3.85 W/m^{2}K obtained from independent air phase simulations is implemented in Eq. 7.2. The volume change of PCM phase is neglected in the computations. Thereby, simplified model accounts natural convection with solid PCM settling in PCM phase. Results from the numerical simulations are validated with optical measurements for PCM melting in cubical cavity.

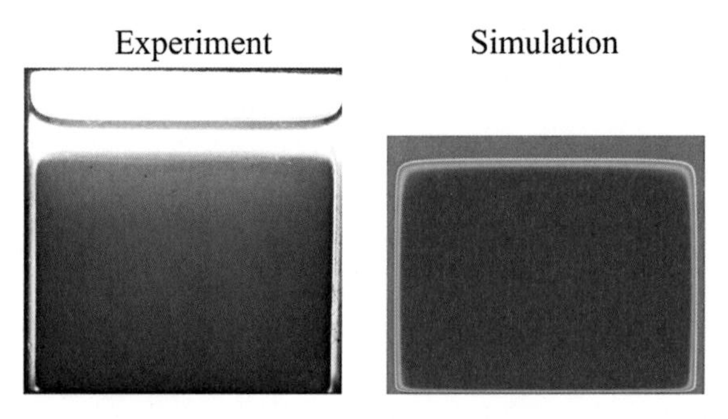

Figure 7.18: Comparison of simulation results with experiments at t = 4 min of the melting process.

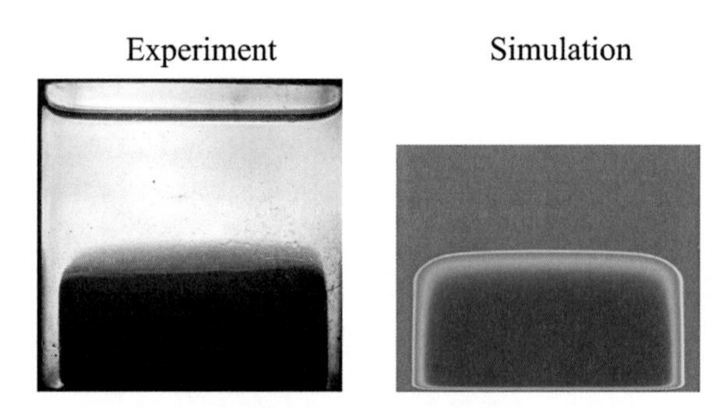

Figure 7.19: Comparison of simulation results with experiments at t = 16 min of the melting process.

Qualitative comparison of numerical simulations has shown a very good acceptance with experiments. The solid PCM height at different times in experiments and simulations also looks similar as seen in Fig. 7.18 and Fig. 7.19. It is also seen that the numerical model estimates solid PCM settling in an excellent manner. Here, a qualitative validation of the numerical model does not provide a descriptive understanding of phase change process. Therefore, the volume fraction of the liquid PCM γ_L from numerical simulations is compared with experiments as seen in Fig. 7.20.

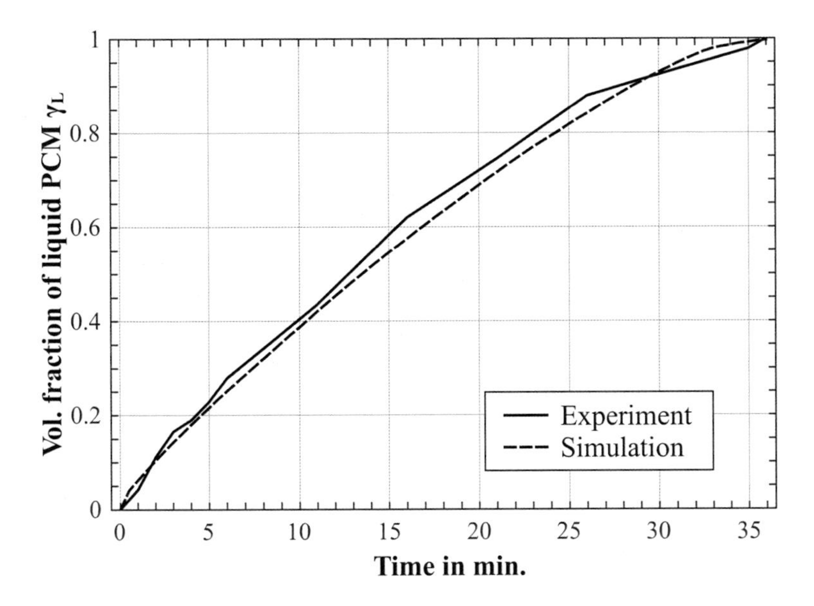

Figure 7.20: Comparison of the liquid PCM vol. fraction from simulation and experiment over t = 36 min.

The melting rate of PCM RT35HC in a cubical cavity is very well represented using the numerical model. Melting of PCM in the capsule is strongly influenced by the liquid natural convection and close contact melting. During melting, natural convection in the liquid PCM enhances the heat transfer in the PCM capsule, whereas close contact melting decreases the thermal resistance of liquid natural convection at the bottom wall of PCM capsule. The movement of phase change boundary towards bottom wall increases the heat transfer and heat flux from PCM bottom wall. Accordingly, the increase in heat flux from bottom wall of PCM capsule is shown in Fig. 7.21. At t = 2 min, the solid PCM settles at the bottom of capsule influencing a rapid shoot up of heat flux value. Thereafter, from close contact melting the total heat flux is dominated by the bottom wall heat flux. The proportion of heat flow from the bottom wall increases consistently with time. By the end of melting process, 97.5 % of heat flow into the capsule enters from the bottom wall. This depicts the importance of close contact melting in a PCM capsule. Accordingly, the accuracy of melting numerical model also depends on the accuracy of settling algorithm to depict close contact melting in a capsule.

As seen in Fig. 7.21, the influence of heat transfer from upper wall of simplified geometry decreases with the progression of time in a melting process. After the solid PCM settles at capsule bottom, thermal resistance due to liquid PCM natural convection increases at the upper wall. A gradual increase in the thermal resistance with increase in the melting column results in a lower heat flow. After t = 5 min, the heat flux from the upper wall falls down to less than 1 % of total heat flux. This value further decreases with further melting of solid PCM and reaches a value less than 0.1 % of the total heat flux by time t = 36 min.

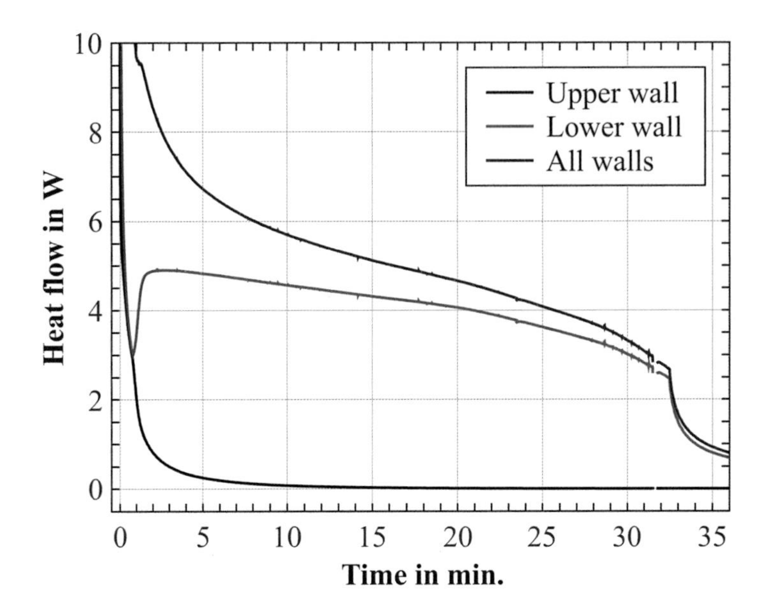

Figure 7.21: Comparison of different wall heat fluxes of the PCM boundaries depicted over the melting time.

7.3.2 Simplified effective conductive model

Numerical modelling and simulation of phase change taking natural convection and close contact melting into account is representative of a real melting process. But, it is arduous to integrate such detailed numerical model for a large scale thermal storage unit application. Due to a large count of PCM capsules, a detailed numerical modelling of thermal storage unit with capsules is both computational expensive and complex. Therefore, some simplified approaches exist to model heat transfer between PCM capsules and HTF in a thermal storage unit. Such simplified approach is developed to implement heat transfer inside a PCM capsule with an effective thermal conductivity. An effective thermal conductivity is relevant for melting of solid PCM. In melting case, the heat transfer inside the PCM capsule is strongly influenced by liquid PCM natural convection and close contact melting. These factors accelerate the melting process through an enhanced heat transfer. Therefore, the effective thermal conductivity takes this enhanced heat transfer due to natural convection and close contact melting into account. In solidification, due to the absence of close contact melting and convection, an effective thermal conductivity is not necessary to represent phase change. Instead a standard thermal conductive process is implemented with the simplified approach. Here, the validation of numerical model supports extension of this approach for other applications and temperature ranges. Melting of RT35HC in a cubical cavity test case is adapted to evaluate the simplified effective thermal conductivity model. A numerical set-up similar to the simplified convective model is employed here. The simplified effective conductive model also takes only the PCM phase into account, to represent phase change of the PCM in a capsule. The secondary air phase in cubical cavity is modelled using a heat transfer coefficient from Fig. 7.16. The numerical set-up and boundary conditions for the test case to validate the effective conductive model is shown in Fig. 7.22.

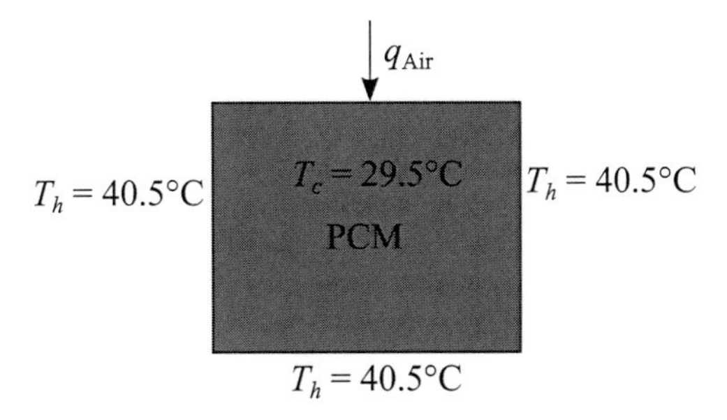

Figure 7.22: Numerical set-up to compute melting of PCM in a cubical cavity with simple effective conductive model.

Detailed close contact melting and natural convection is not modelled here. Therefore, an increased heat flow from capsule bottom wall is not observed. Velocity and pressure are also not solved here. Instead, an enhanced melting rate from close contact melting is accounted using constants c and n, in the numerical model. These constants in Eq. 5.74 estimate the heat transfer during melting of solid PCM. In this model, effective thermal conductivity is computed at every time step, which depends on the increase and decrease of melt column in an enclosure. An increase in melting column influence of rise of natural convection in melting process. Such influence is implemented in this numerical model with a variable effective thermal conductivity. The increase in the effective thermal conductivity as seen in Fig. 7.23 is influenced by constants c, n and Ra_L. The effective thermal conductivity estimated for the melting and settling of solid RT35HC in a cubical cavity is

$$\kappa_{eff,\,L} = \kappa_L \cdot 0.245 \cdot [\mathrm{Ra}_L \cdot V(\gamma_L)]^{0.245}. \tag{7.3}$$

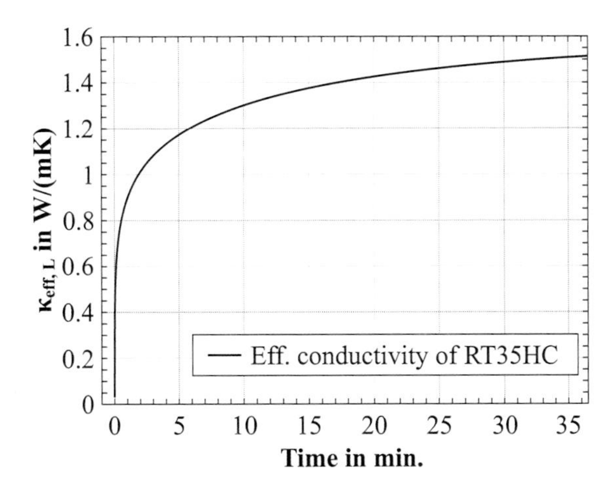

Figure 7.23: Effective thermal conductivity increment in melting process of RT35HC.

The melting rate from the effective conductive model is validated with experiments as shown in Fig. 7.24

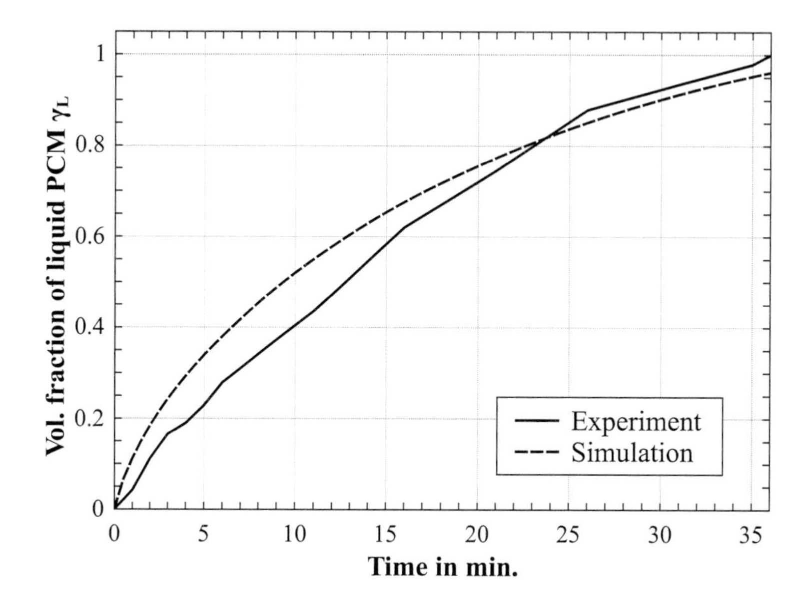

Figure 7.24: Comparison of the liquid PCM vol. fraction from simulation and experiment over t = 36 min.

7.3.3 Role of natural convection and solid settling in melting of salt hydrate inside a cylindrical enclosure

Simplified numerical models for melting of encapsulated PCM are validated with experiments. These numerical models have shown a very good acceptance with experiments along with a good estimate of the physical process. This asserts the correctness of the numerical model to estimate solid-liquid phase change. Therefore, simplified effective numerical model is employed to depict melting of salt hydrate, $MgCl_2 \cdot 6H_2O$. The thermo-physical properties of $MgCl_2 \cdot 6H_2O$ are described in Tab. 7.2.

Table 7.2: Thermo-physical properties of $MgCl_2 \cdot 6H_2O$ [211].

Physical property	Symbol	Units	Value
Melting temperature	T_m	°C	115
Latent heat	L	J/kg	166900
Specific heat capacity in solid phase	c_s	J/kgK	1830 @ 100 °C
Specific heat capacity in liquid phase	c_l	J/kgK	2570 @ 120 °C
Thermal conductivity in solid phase	λ_s	W/mK	0.7 @ 110 °C
Thermal conductivity in liquid phase	λ_l	W/mK	0.63 @ 120 °C
Solid phase density	ρ_s	kg/m^3	1595 @ 20 °C
Liquid phase density	ρ_l	kg/m^3	1456 @ 120 °C

For this, constants for effective thermal conductivity are adapted to measure the influence of natural convection and close contact melting during phase change.

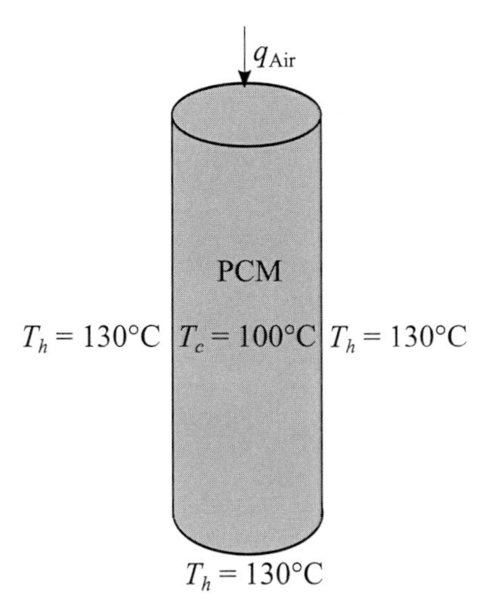

Figure 7.25: Numerical set-up to compute melting of $MgCl_2 \cdot 6H_2O$ in a cylindrical enclosure with simple effective conductive model.

The experimental set-up as seen in Fig. 6.7 is deployed to compute melting of $MgCl_2 \cdot 6H_2O$ in a cylindrical enclosure. The boundary conditions for the numerical set-up is depicted in Fig. 7.25. Due to its relevance of the application in a large scale thermal storage unit, a cylindrical geometry has been chosen. For this test case, salt hydrate $MgCl_2 \cdot 6H_2O$ as PCM is also adapted for the same purpose.

Natural convection without solid settling

In this section, the experimental set-up is developed to investigate temperature distribution in PCM during melting. Due to an immersed thermocouple at PCM centre, the solid PCM remains fixed during melting. This obstructs solid PCM settling in the cylindrical enclosure during solid-liquid phase change. Also, due to a high operating temperature range alongside PCM opaqueness, optical measurement cannot be employed. Thereby, heat transfer during the melting of $MgCl_2 \cdot 6H_2O$ is investigated through thermometry. Coefficients c and n are investigated to estimate the influence of pure natural convection from effective thermal conductivity, which is compared with experiments. The range of values c and n are approximately obtained from the correlations shown in Tab. 5.1. This calibrated effective thermal conductivity correlation depicting the melting of $MgCl_2 \cdot 6H_2O$ in a cylindrical enclosure is written as

$$\kappa_{eff,\,L} = \kappa_L \cdot 0.09 \cdot [\mathrm{Ra}_L \cdot V(\gamma_L)]^{0.23} \,. \tag{7.4}$$

Melting of $MgCl_2 \cdot 6H_2O$ with an effective thermal conductivity model is computed with a correlation from Eq. 7.4. The results from numerical simulation are compared with experiments.

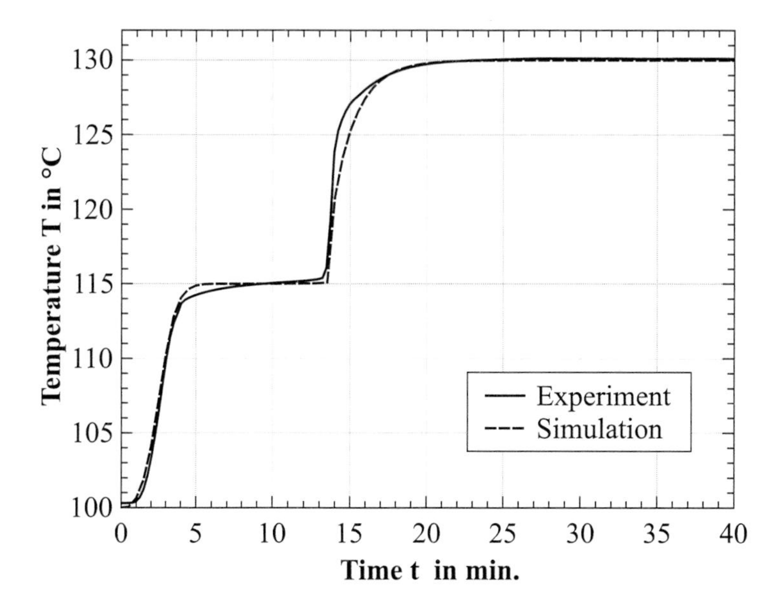

Figure 7.26: Temperature comparison between simulation results from the effective thermal conductivity model and the measured temperature for experiments at the cylindrical enclosure centre.

As seen in Fig. 7.26, effects of natural convection in a melting process can can be very well estimated. Where, the simplified effective thermal conductivity followed a correlation in Eq. 7.4.

Natural convection with solid settling

To estimate the influence of solid settling and natural convection, the effective thermal conductivity correlation is compared with the results from simple convective model. The effective parameters are adapted for effective conductivity correlation to estimate the influence of solid settling and natural convection for $MgCl_2 \cdot 6H_2O$ in a cylindrical enclosure. These effective thermal conductivity parameters like c and n from Eq. 5.72 are evaluated, where the correlation yields

$$\kappa_{eff,\,L} = \kappa_L \cdot 0.08 \cdot \left[\mathrm{Ra}_L \cdot V(\gamma_L)\right]^{0.2}. \tag{7.5}$$

For this, the melting rate from effective conductive model is compared with the simplified convective model. As seen in Fig. 7.27, the melting rate of $MgCl_2 \cdot 6H_2O$ inside cylindrical enclosure is very well estimated from Eq. 7.5. Here, the reduced values of c and n are observed from the cylindrical configuration. Due to a low aspect ratio of cylindrical geometry, the surface area of bottom wall contributes only to 5 % of the total cylinder surface area. Therefore, major heat transfer during phase change is contributed from the longitudinal side walls of cylinder. In such case, the heat transfer in a melting process is delayed by time t = 3 min for a solid settling case. Accordingly, a lower effective thermal conductivity is adapted for heat transfer inside a capsule with solid settling. The lower effective thermal conductivity illustrates delayed melting and reduced heat transfer with solid settling.

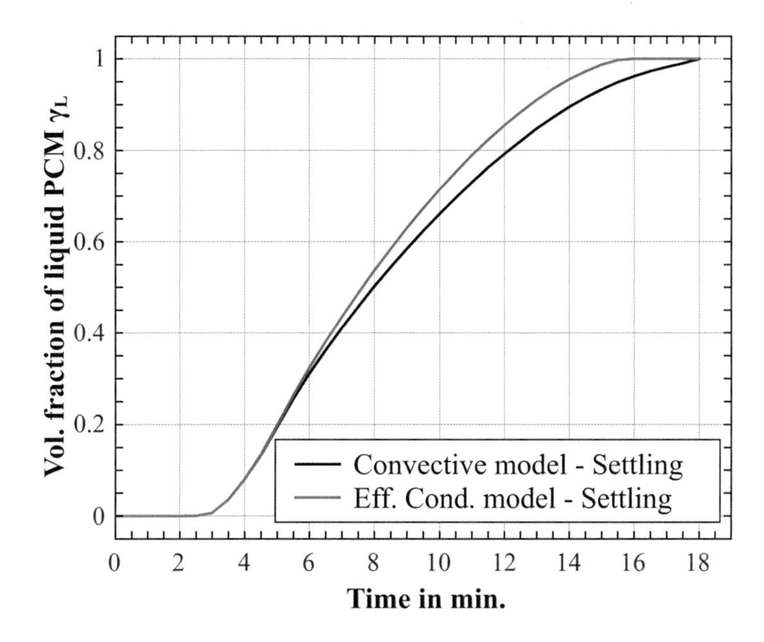

Figure 7.27: Comparison of melting rate from the effective conductive model and simplified convective model.

For the use in the laboratory scale thermal storage unit as seen in Fig. 6.15, a similar cylindrical geometry with low aspect ratio is employed. The surface area of the bottom wall for this cylinder lies also close to 5 % of total surface area, which makes it similar. Therefore, the validated correlation from Eq. 7.5 is employed to represent effective thermal conductivity for a melting process of $MgCl_2 \cdot 6H_2O$ influenced by natural convection and solid settling.

7.4 Conjugate heat transfer (CHT) model

In this work, solid-liquid phase change inside a PCM capsule is computed from simplified to detailed scale. In these models, thermo-fluidic behaviour of HTF is modelled with a boundary condition. Therefore, the developed conjugate heat transfer (CHT) model computes heat transfer between HTF, capsule wall and PCM. The heat transfer between these different materials is modelled through coupled overlapping interface approach. Accordingly, the numerical model takes flow of HTF, phase change in PCM and thermal conductivity of capsule wall into account. The results from the numerical simulations are validated with experimental findings. Here, the experimental set-up consists of three cylindrical PCM capsules made of glass immersed in a rectangular container. The fluid flow around the capsules is accommodated as seen in Fig. 6.9 and Fig. 6.11. A similar representation of the experiment is implemented in the computational domain. The boundary conditions on the numerical set-up are implemented as shown in Fig. 7.28.

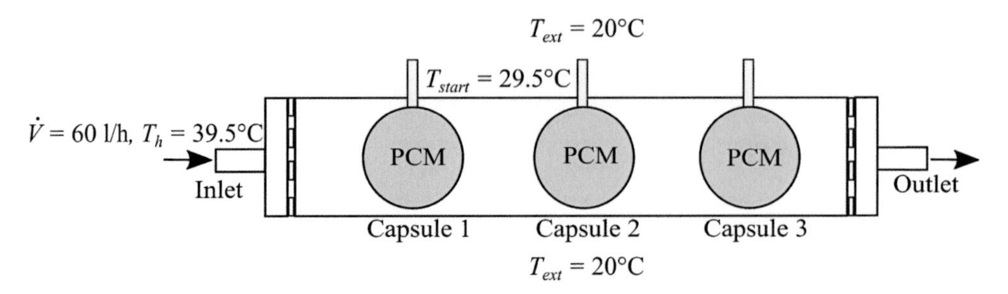

Figure 7.28: Numerical set-up and boundary conditions to validate the CHT model with experiments.

Due to low permissible operating temperatures, water is used as a HTF in the experiments. Accordingly, temperature dependent density of water is implemented in the numerical model. Initially, a temperature condition T_{start} = 29.5 °C on the coupled computational domain. HTF with a temperature of T_h = 39.5 °C is flow around the capsules with a volumetric flow rate of 60 l/h. Similar to experiments, in numerical simulations solid PCM in capsule 1 and capsule 3 experience close contact melting. Whereas, the settling of solid PCM in capsule 2 is restricted. Temperature measured at different positions of container are compared with numerical results.

Validating the melting of PCM

The solid PCM surface area including the melt rate from numerical simulations are compared with experimental results. The qualitative validation of numerical model is shown in Fig. 7.29.

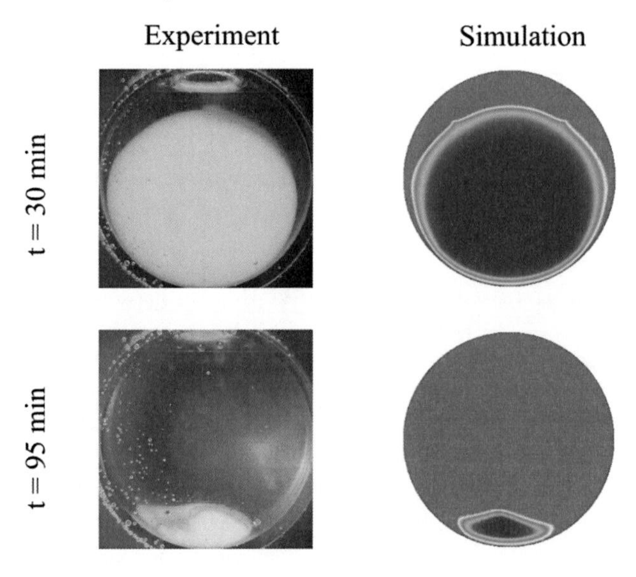

Figure 7.29: Qualitative comparison PCM melting in the capsule 1 between numerical simulations and experiments.

The comparison of solid PCM in capsule 1 from numerical simulations with experiments has shown a very good acceptance. The numerical model predicts the solid PCM dimensions in a melting process very well at different times. A small deviation in Fig. 7.30 is influenced by the solid PCM rotatory motion in experiments.

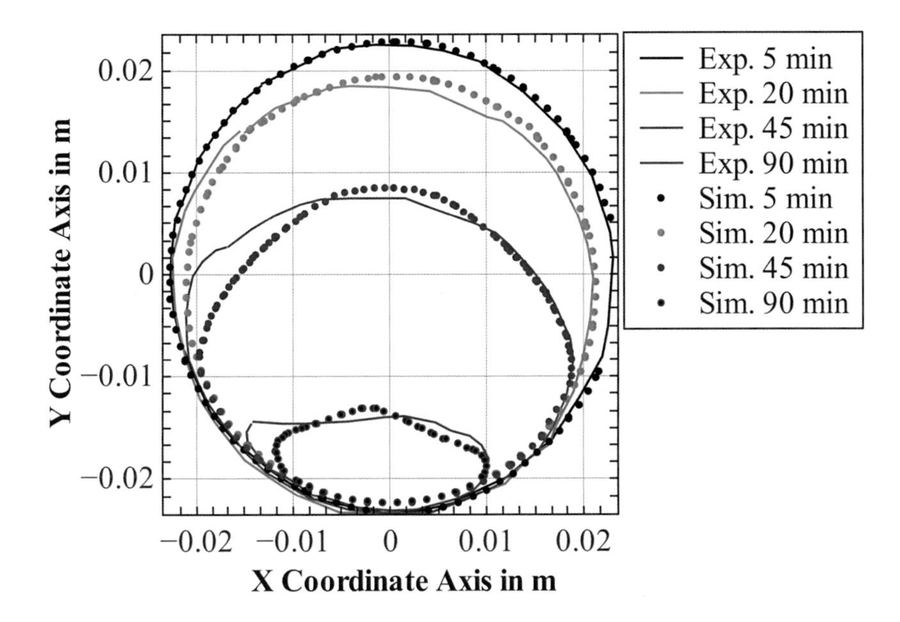

Figure 7.30: Comparison of solid-liquid interface from experiment and simulation at different times.

For a quantitative comparison, the melting rate of the PCM is compared with experiments. In experiments, the solid PCM melting is evaluated from 2D photographs of the capsule. The PCM melt from the captured 2D photograph is calculated by excluding the solid PCM domain. This evaluation method works very well for sidewards heated PCM capsule set-up. But, the cylindrical PCM capsule in this experimental set-up is heated from all sides. Therefore, a solid PCM surface area from experiments is not sufficient to computed the volume fraction of liquid PCM. Instead a solid topological volume is required to compare the volume fraction in numerical simulations with experiments. Therefore, the solid PCM volume from 2D photographs are interpolated over a 3D domain in experiments. This obtained volume of 3D solid PCM is evaluated to calculate the volumetric melt rate of the solid PCM. This adapted volume fraction of liquid PCM is compared with numerical simulations in Fig. 7.31. Here, an unsteady melting rate from 2D photographs is also shown. Numerical results have shown a close agreement with the 3D liquid PCM volume fraction in experiments. The numerical model predicts the heat transfer between HTF and PCM capsule 1 including HTF flow very well as seen in Fig. 7.31.

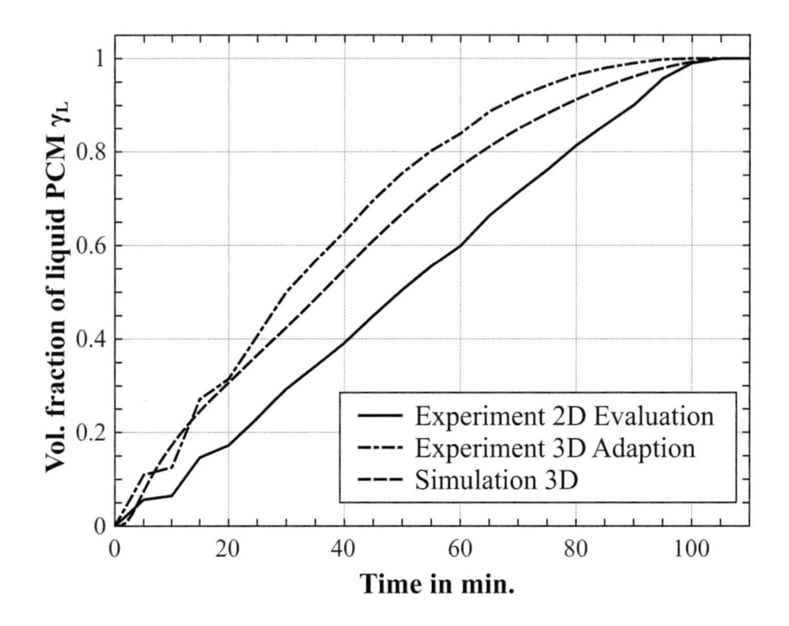

Figure 7.31: Melt rate of the PCM RT35HC in the capsule 1 from the numerical simulations compared with the 2D and 3D evaluation of melt rates from the experiments.

Validating the thermal distribution in HTF

During charging, PCM capsules in the container gain thermal energy from HTF. In experiments, temperature measured at different positions of flow is compared with temperature distribution in numerical simulations. Such a comparison can be observed in Fig. 7.32, which explains the accuracy of numerical simulations. It is noticed that experimental and numerical results differ with a thin offset. The deviation arises from a small experimental thermocouple error, which is estimated to be around 1.5 °C for K type thermocouple. This error in experiments is clearly evident due to its consistency in temperature curves for TC01 and TC03. Therefore, this estimate error of 1 °C can be tolerated, which show a very good acceptance of numerical simulations with experiments. From the present model, forced convection and coupled heat transfer in HTF are represented accurately. Heat flow into different PCM capsules in the set-up can be compared to understand the influence of close contact melting. Due to the fixed solid, capsule 2 experiences different heat transfer profile as capsule 1 and 3. A small bend in the heat flow curve is noted after solid PCM settling at capsule bottom. Such a bend is observed in Fig. 7.33 for capsule 1 and capsule 3 at t = 20 min. Despite the lateral distance between the capsule 1 and capsule 3, the transient heat flow into the capsules look very similar in Fig. 7.33. However, the solid PCM inside the capsule 1 melts sooner than in capsule 3, whose solid PCM melts sooner than PCM in capsule 2.

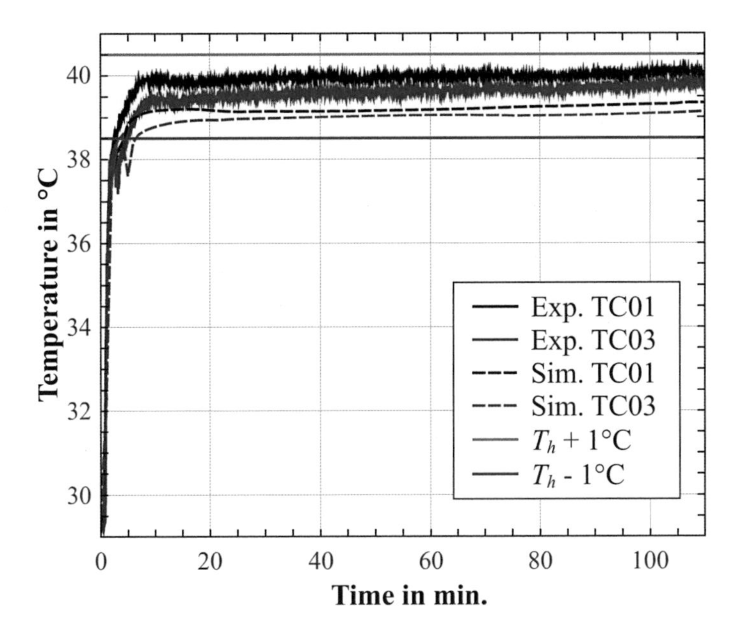

Figure 7.32: Comparison of temperature plots between experiments and numerical simulations at thermocouples TC01 and TC03 measured inside the HTF flow.

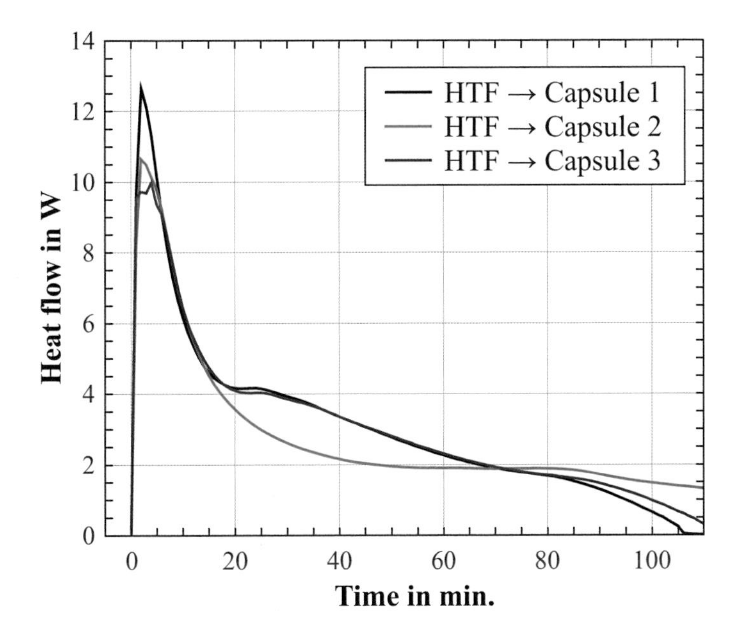

Figure 7.33: Comparison of the heat flow rate into the PCM capsules from the HTF.

Heat flow into PCM capsule experiences a thermal resistance from PCM capsule wall. Due to a decent thermal conductivity of glass, very small thermal resistance is offered by the capsule wall. An actual heat flow into PCM is calculated from heat flow into PCM capsule and thermal resistance offered from capsule wall. The heat flow into PCM responsible for the melting of solid PCM is shown in Fig. 7.34.

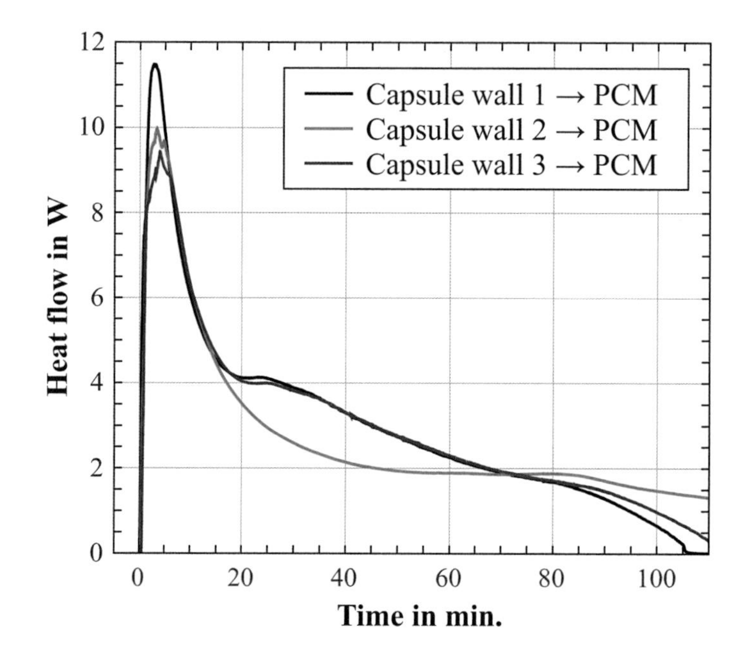

Figure 7.34: Comparison of the heat flow rate into PCM from respective capsule wall.

The influence of wall thermal resistance is evidential from the peak power in heat flow curve in Fig. 7.33 and Fig. 7.34. The difference in peak value is caused by thermal resistance. In Fig. 7.34, the influence of close contact melting is also clearly evidential, which results in a faster melting of PCM with solid settling.

Density change in the HTF due to temperature difference forms a layered flow, which is observed in temperature profiles of experiments and numerical simulations. Due to low operating velocity, the denser HTF settles at the bottom of container. This behaviour is observed in initial instances of charging as seen in Fig. 7.35. Solid PCM settling in capsule 1 and capsule 3 is depicted in Fig. 7.35. Temperature distribution in set-up at times t = 2 min and t = 55 min is also shown in Fig. 7.35. Due to a very small inlet diameter, high velocity is observed at the inlet. However, it falls down to a very low value after crossing the perforated plate. Here, the perforated plate enables a homogeneous distribution of HTF flow in the container as shown in Fig. 7.36.

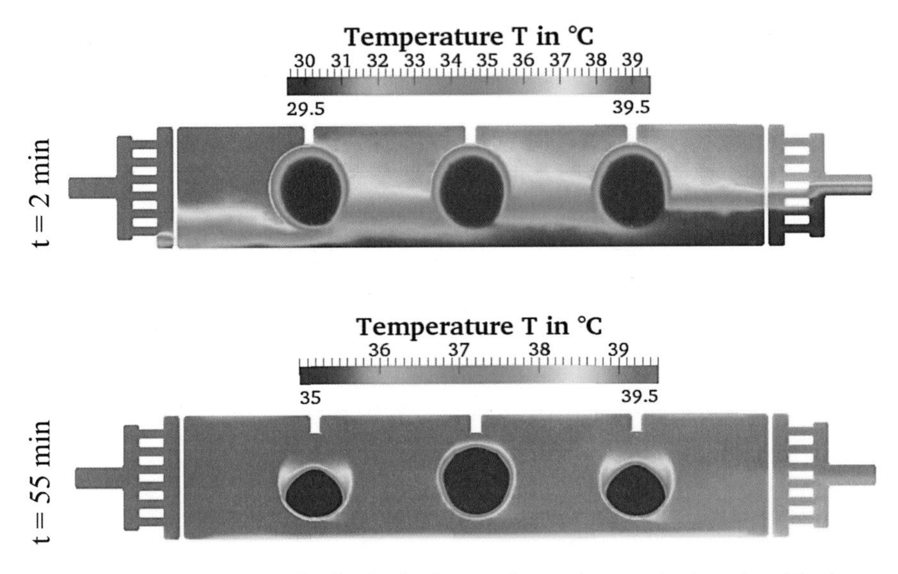

Figure 7.35: Temperature distribution in the container at times t = 2 min and t = 55 min.

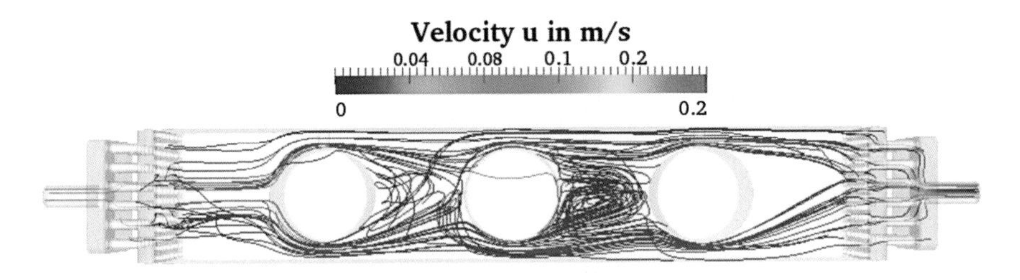

Figure 7.36: 3D flow streamlines of HTF in the container after achieving a quasi-steady flow distribution at time t = 110 min.

Different operating conditions and parameters can be studied with the present numerical model. A detailed modelling of conjugate heat transfer inside a thermal storage unit enables better understanding of charging and discharging processes. However, the conjugate heat transfer model is limited due to its computational effort.

7.5 Local thermal non equilibrium (LTNE) model

Due to a large computational effort required by the conjugate heat transfer model, a simplified LTNE model for a whole LHTES unit is developed here. With this simplified model, a local thermal non-equilibrium between PCM capsules and HTF inside a storage unit is modelled. Accordingly, the LTNE model accounts the unsteady heat transfer process between PCM capsules and HTF. Two different material domains are implemented in this approach. Charging and discharging of macro-encapsulated LHTES unit is computed from the numerical model, which is compared with experimental results. In a complete macro-encapsulated LHTES unit, just the packed bed with HTF is modelled in the numerical simulations. The numerical set-up and boundary conditions of the experimental set-up are shown in Fig. 7.37.

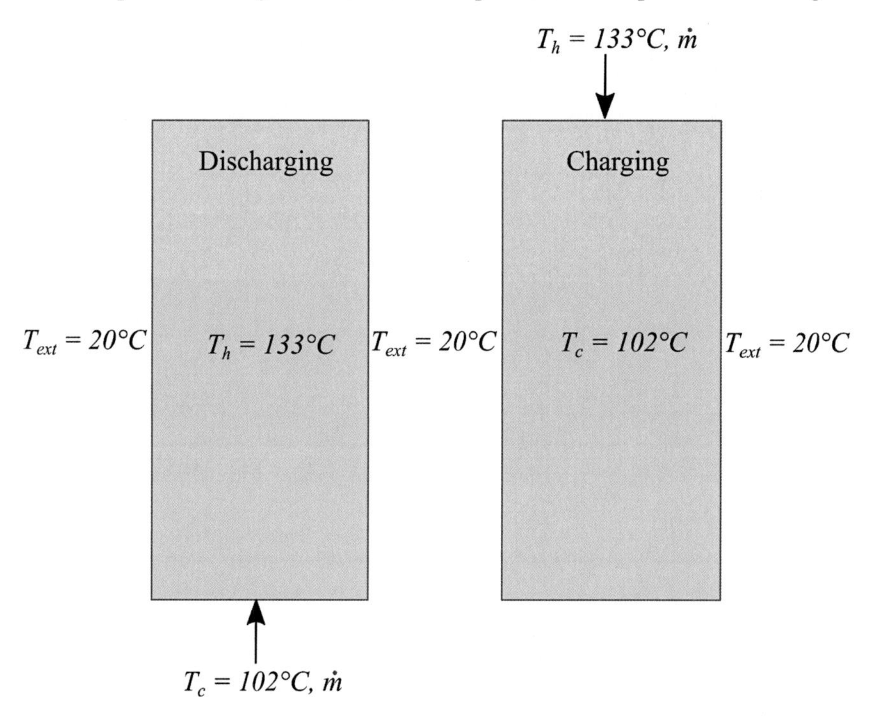

Figure 7.37: Schematic representation of numerical set-up and boundary conditions for latent heat thermal storage units.

LHTES unit with 72 cylindrical capsules having an external diameter of 40 mm and length 250 mm are arranged in two columns, whose simulation parameters are shown in Tab. 7.3 and Tab. 7.4. The materials properties of the HTF are shown in Tab. 7.3. For different mass flow rates of HTF, the numerical model is computed and validated with experiments. Heat losses in charging and discharging processes of latent heat thermal storage units is implemented with an overall heat transfer coefficient. The charging and discharging of laboratory scale LHTES is operated at different temperature ranges of 102 °C and 133 °C. The volume fraction of HTF is implemented as the porosity of container. A detailed representation of thermal storage unit and capsule form are found in Fig. 6.16 and Fig. 6.15. The capsule thickness configuration and type of encapsulation is shown in Fig. 6.15. The cylindrical capsules are filled with the salt hydrate $MgCl_2 \cdot 6H_2O$, whose material properties are shown in Tab. 7.2. The packed bed of this LHTES unit has a global porosity equal to 0.34197. PCM capsules with 250 mm length are placed in two sets on one another, which have a packing length of 500 mm. Here, an active porosity of the packed bed inside a packing length of 0.25 m $\times$ 0.25 m $\times$ 0.5 m is 0.2761. Because, in the LHTES unit only HTF exists above this packing length. This is implemented in the numerical simulations with a length dependent porosity. Where, the porosity of LHTES unit takes a value of 0.2761 inside the packing length and 1 for the rest of container.

Table 7.3: Material properties of HTF.

Material properties	units	Value
Density (ρ) @ 20 °C	kg/m^3	1034
Viscosity (μ)	Pa ·s	0.011374
Specific heat capacity (c)	J/kg·K	2230
Thermal conductivity (κ)	W/m·K	0.159 .

Similarly, the material properties of the capsule wall are shown in Tab. 7.4.

Table 7.4: Material properties of PCM capsule wall.

Material properties	units	Value
Density (ρ) @ 20 °C	kg/m^3	2700
Specific heat capacity (c)	J/kg.K	917
Thermal conductivity (κ)	W/m.K	210.

7.5.1 Discharging of laboratory scale latent heat thermal storage unit

Thermal discharging of a macro-encapsulated LHTES unit is computed with the developed LTNE model. The results from the numerical simulations are validated with experiments, which depicts the accuracy of numerical model. Thermal discharging process for operating mass flow rates 5 kg/min and 10 kg/min is computed here. During thermal discharging, the PCM inside the capsules solidify from liquid to solid. In this case, the effects of liquid natural convection and close contact melting can be neglected. Therefore, an effective thermal conductivity is not necessary to represent the phase change in a PCM capsule. In this model, an overall heat transfer coefficient of a PCM capsule is computed and updated in every time step. The overall heat transfer coefficient computes the heat transfer between a single PCM capsule and HTF. Outlet temperature plot during thermal discharging of the present macro-encapsulated LHTES unit for a temperature difference of ΔT = 30 K and mass flow rate of 5 kg/min is depicted in Fig. 7.38. Here, inlet temperature is also shown to depict the steady state of discharging process.

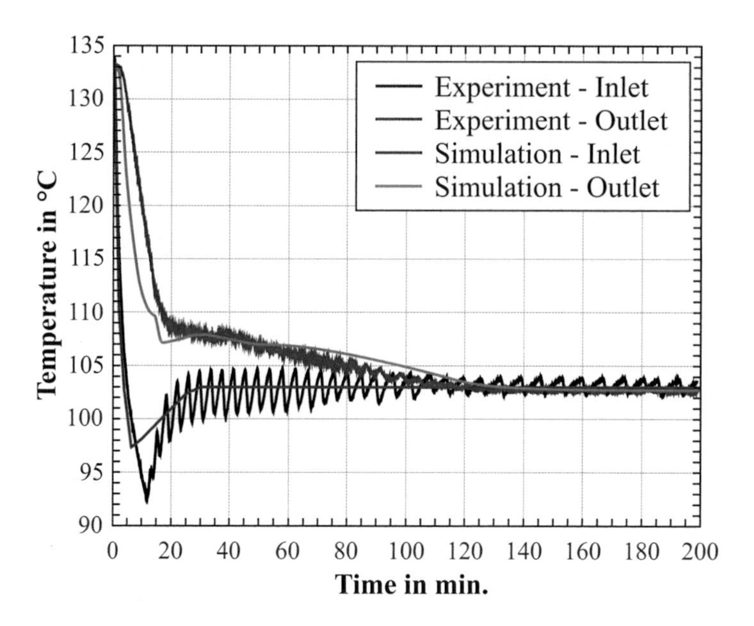

Figure 7.38: Comparison of the inlet and outlet temperature profiles of experiments and simulations for mass flow rate 5 kg/min within the discharging time.

Similarly, outlet and inlet temperature plot for thermal discharging of latent heat thermal storage unit with operating mass flow rate 10 kg/min is shown in Fig. 7.39.

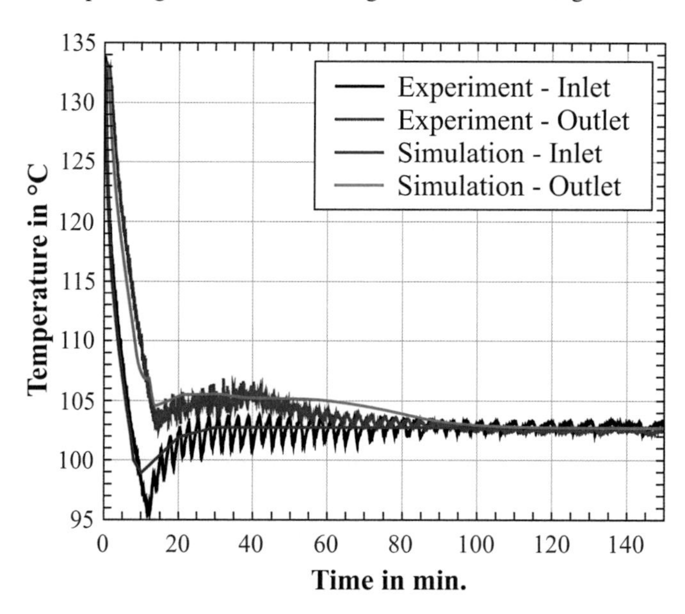

Figure 7.39: Comparison of the inlet and outlet temperature profiles of experiments and simulations for mass flow rate 10 kg/min within the discharging time.

Numerical results from the LTNE model have shown good acceptance with experiments. Unsteady heat transfer including with the fluid flow in laboratory scale LHTES unit is very well presented using the numerical model. Fluctuations in the temperature profiles from experiments seen in Fig. 7.38 and Fig. 7.39 are caused from a coarse regulation of thermostat. Here, temperature regulator in the experimental set-up produces fluctuating temperatures into the storage unit. This transient inlet temperature profile is implemented in simulations with an unsteady temperature boundary condition at the inlet. Temporal mean of fluctuating temperature at regular intervals is calculated to represent this unsteady inlet temperature profile. For a mass flow rate of 5 kg/min, a quasi steady state of discharging process is achieved at t = 132 min in the experiments and time t = 134 min in numerical simulations as seen in Fig. 7.38. As seen in Fig. 7.39, for mass flow rate of 10 kg/min, a similar quasi steady state of discharging process is achieved at time t = 117 min in the experiments and at time t = 119 min in numerical simulations. For a mass flow rate of 5 kg/min, thermal power of the discharging process is shown in Fig. 7.40.

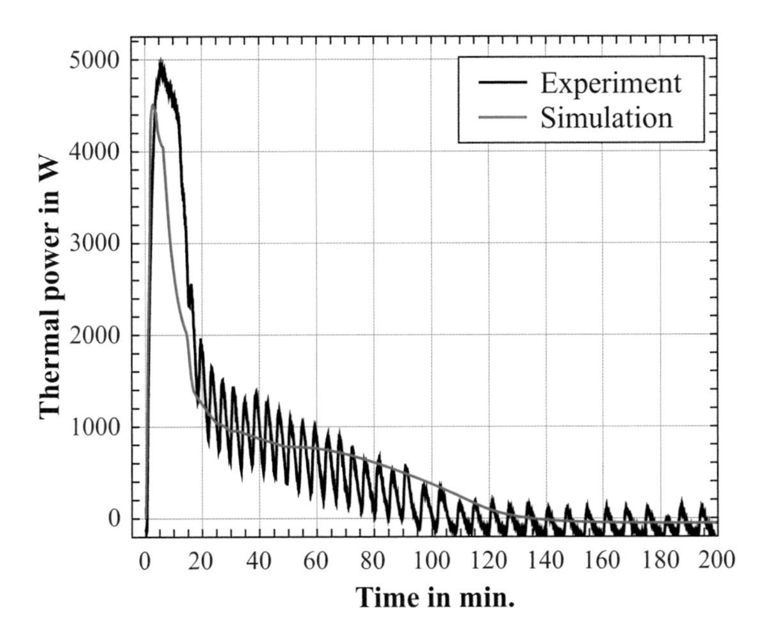

Figure 7.40: Discharging power of thermal storage unit from experiments and simulation for a mass flow rate 5 kg/min.

A comparison of the storage discharging power from numerical simulations has shown a very good acceptance with experiments. A minute difference in thermal power peak is observed in Fig. 7.40 between experiments and simulation. This is caused due to a deep fall of temperature value in experiments at time t = 12 min as shown in Fig. 7.38. Due to an unsteady temporal average of inlet temperature, this sharp change is not implemented in the numerical simulations. A similar discharging power plot can also be represented for the operating condition 10 kg/min in Fig. 7.41.

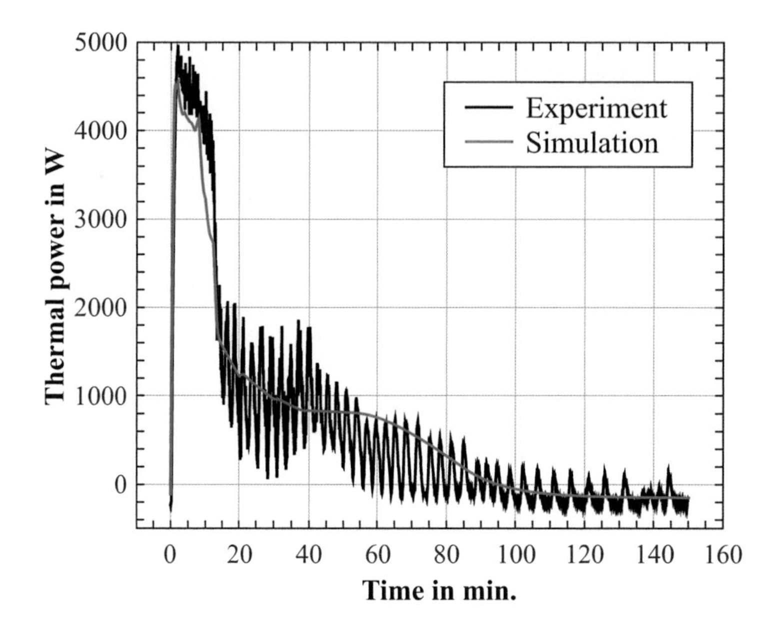

Figure 7.41: Discharging power of the thermal storage unit from experiments and simulation for a mass flow rate 10 kg/min.

As seen from Fig. 7.41, the numerical model predicts discharging process of present storage unit very well also for an operating mass flow rate 10 kg/min.

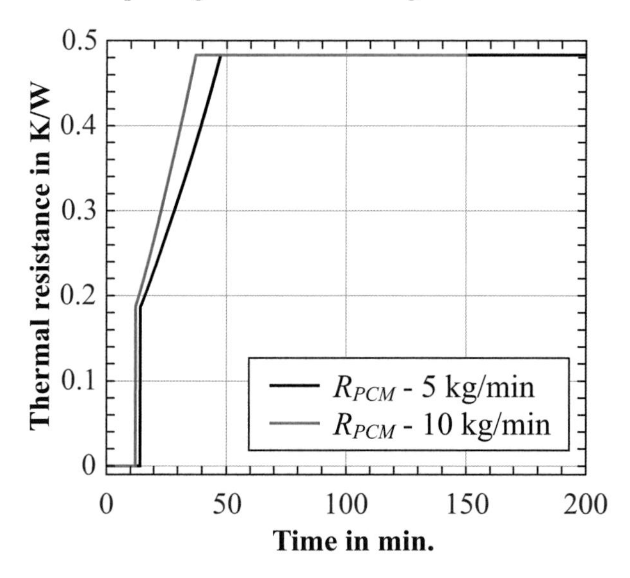

Figure 7.42: Unsteady thermal resistance offered by PCM during discharging process.

In the numerical model, thermal resistance offered by the phase change column of PCM is modelled in an unsteady manner. The thermal resistance is calculated from the phase change volume fraction at every time step. Here, the overall heat transfer coefficient is computed from the sum of thermal resistances offered by the capsule wall, external fluid convection and PCM melt column as explained in Eq. 5.116. Thermal resistance offered by the phase change column of PCM is set to zero till 10 % of PCM inside the PCM capsule experiences phase change. Thereafter, the thermal resistance in phase change column is calculated, which contributes to the total thermal resistance. In the numerical model, thermal resistance from PCM phase change column grows with the increase in phase change boundary. In the numerical model, it takes a constant value when 75 % of total PCM volume in the thermal storage attains phase change. This unsteady thermal resistance offered by PCM phase change column is shown in Fig. 7.42. The difference in PCM thermal resistance between mass flow rates 5 kg/min and 10 kg/min as in Fig. 7.42 is due to a faster phase change of PCM for a higher mass flow rate.

From numerical simulations, mean thermal discharge power of the present laboratory scale LHTES unit is 890 W for mass flow rate 5 kg/min and 1115 W for mass flow rate 10 kg/min. Here, the mean thermal discharge power is calculated by neglecting minute power contributions lesser than 1 % of maximum peak.

7.5.2 Charging of laboratory scale latent heat thermal storage unit

In this section, thermal charging of the present laboratory scale LHTES unit is computed with LTNE model. Here, the results from numerical simulations are further validated with experiments. Operating conditions like mass flow rate and temperature differences are employed similar to the discharging process. During charging, the solid PCM in PCM capsules stores thermal energy from the HTF through phase change. Melting of PCM inside the PCM capsules occurs during a charging process. In the numerical model, melting of PCM is implemented using the effective thermal conductivity approach. Effects like natural convection in liquid PCM phase and solid PCM settling inside a PCM capsule are adapted with an effective thermal conductivity here. In the numerical model, effective thermal conductivity value increases with natural convection rise and close contact melting inside the capsule. Effective thermal conductivity value is computed from the volume fraction of fluid PCM inside a capsule. Previously in this work, an effective thermal conductivity correlation is developed for a cylindrical capsule filled with $MgCl_2 \cdot 6H_2O$. This correlation is adapted for a temperature difference 30 K, which is validated with experiments in Fig. 7.27. This fitted correlation for melting and settling of $MgCl_2 \cdot 6H_2O$ in a cylindrical enclosure is written as

$$\kappa_{eff,\,L} = \kappa_L \cdot 0.08 \cdot [\mathrm{Ra}_L \cdot V(\gamma_L)]^{0.2} \,. \tag{7.6}$$

Due to similar geometrical configuration and operating conditions, this correlation for natural convection and solid settling is also relevant here. Therefore, it is employed to represent an effective thermal conductivity of PCM in a charging process. Whereby, the influence of natural convection and close contact melting inside a cylindrical PCM capsule is taken into account. For a charging process with temperature difference ΔT = 30 K and operating mass flow rate of 5 kg/min, inlet and outlet temperature plot of the present laboratory scale LHTES unit is shown in Fig. 7.43.

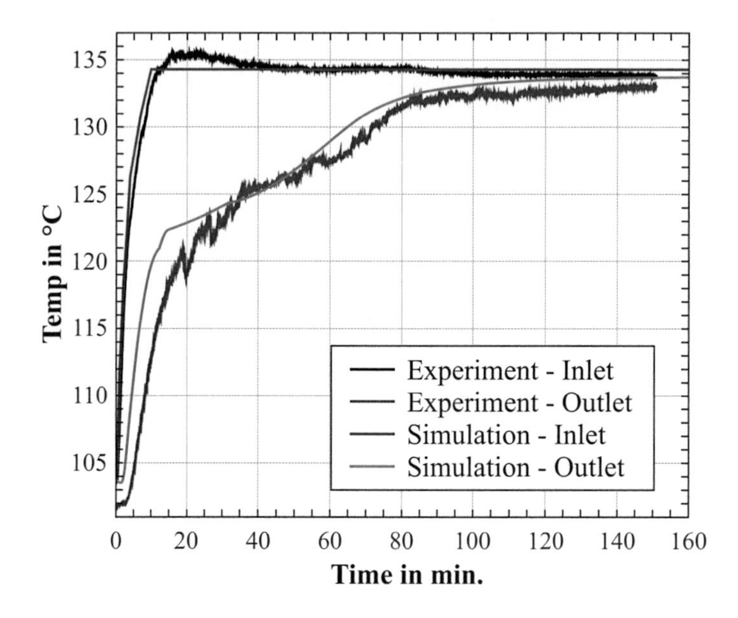

Figure 7.43: Comparison of inlet and outlet temperature profiles of experiments and simulations for mass flow rate 5 kg/min within the charging time.

Similarly, outlet temperature plot during thermal charging of laboratory scale LHTES unit with mass flow rate of 10 kg/min is depicted in Fig. 7.44.

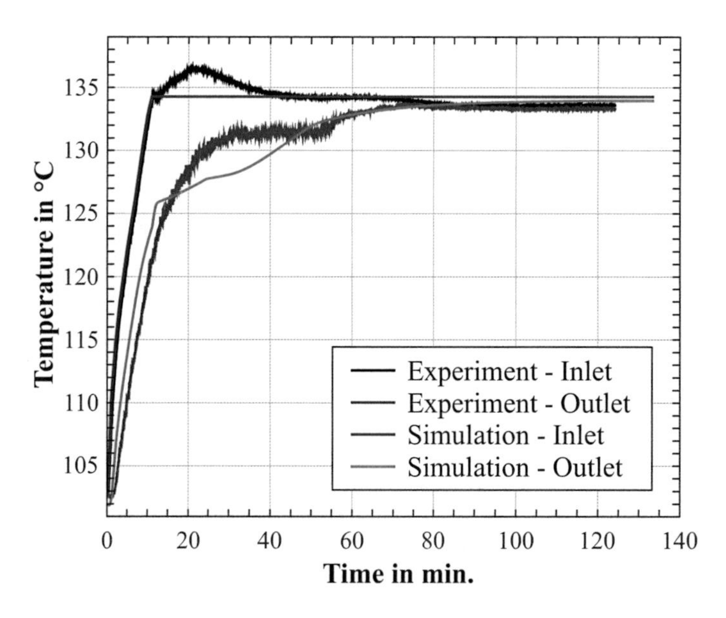

Figure 7.44: Comparison of the inlet and outlet temperature profiles of experiments and simulations for mass flow rate 10 kg/min within the charging time.

A comparison of numerical results with experiments have shown very good acceptance. However, a small deviation is evident from a fluctuating inlet temperature profile in experiments. These unsteady temperature fluctuations are modelled as temporal mean in the numerical simulations. At some instances, a jump in the inlet temperature value from experiments has shown a very large difference with the temporal mean value in numerical simulations. These deviations are also seen in the storage power output from Fig. 7.45 and Fig. 7.46. An unsteady treatment of PCM thermal resistance in the numerical model can be noticed in the form of bends in the temperature plots as seen in Fig. 7.43 and Fig. 7.44. For operating mass flow rate 5 kg/min, thermal power during the charging of laboratory scale LHTES unit is shown in Fig. 7.45.

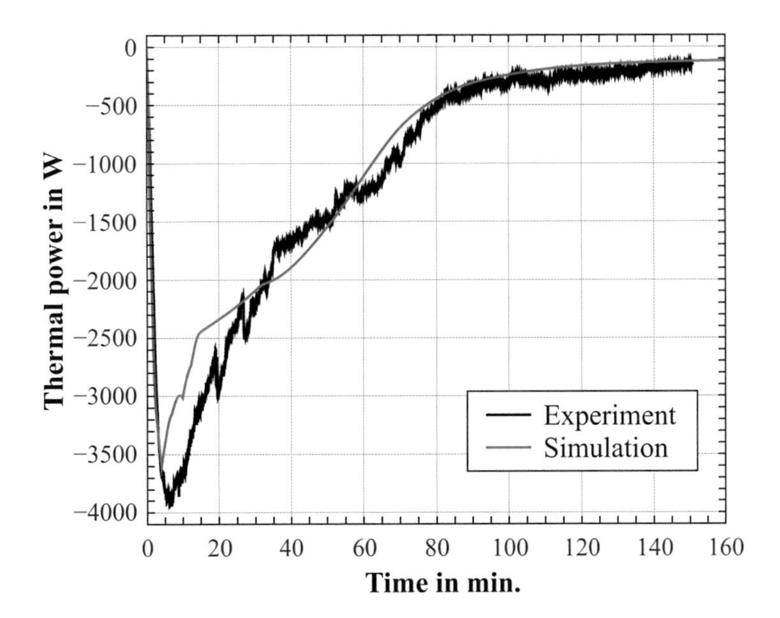

Figure 7.45: Charging power of the thermal storage unit from experiments and simulation for a mass flow rate 5 kg/min.

Numerical simulations have predicted the charging of the present storage unit very similar to experiments. A typical phase change profile observed between time t = 30 min and time t = 100 min is similar in both experiments and simulations. Nevertheless, a deviation of thermal power estimate from numerical simulations with experiments is influenced from inlet temperature jump in experiments. For operating mass flow rate 10 kg/min, thermal power during the charging of a storage unit is shown in Fig. 7.46.

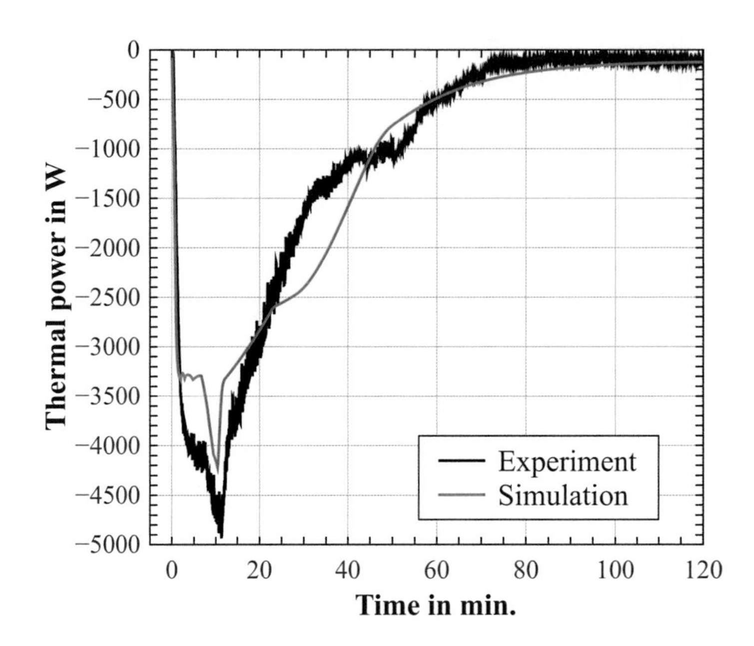

Figure 7.46: Charging power of thermal storage unit from experiments and simulation for a mass flow rate 10 kg/min.

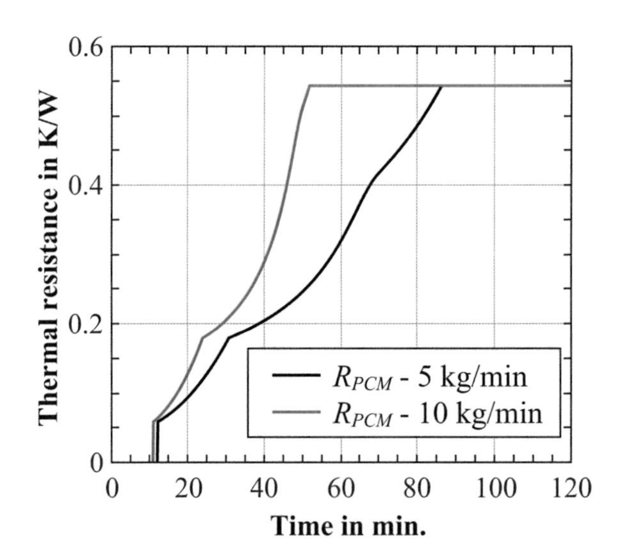

Figure 7.47: Unsteady thermal resistance offered by PCM during discharging process.

Thermal resistance from the melt column is neglected at the initial instances of charging process. Thermal resistance from the melt column is computed only after 10 % of total PCM volume is melt. Thereafter, thermal resistance from PCM melt column rises with the increase in melt column. The total resistance offered by PCM melt column on the overall heat transfer coefficient is a combination of lateral and radial thermal resistances of PCM melt column. In a discharging process, the growth in PCM thermal resistance is neglected when 75 % of PCM volume experiences phase change. But, in a charging process from close contact melting and natural convection, this cannot be neglected so early. Instead, PCM resistance takes a constant value only when $[V(\gamma_L)]^{0.2}$ reaches 75 % of PCM volume in charging process. Here, due to a greater distance between the solid-liquid interface and PCM capsule wall, influence of thermal resistance rise from PCM column on heat transfer can be neglected. Thermal resistance offered by PCM melt column against the heat transfer between PCM and HTF behaves differently in charging process as depicted in Fig. 7.47.

The mean thermal charging power of the present LHTES unit is 1025 W for mass flow rate 5 kg/min and 1200 W for mass flow rate 10 kg/min. Here, the mean thermal charging power is computed by neglecting very small power contributions lesser than 1 % of maximum peak power.

Energy balance

During thermal charging and discharging of storage unit, thermal energy is stored and delivered. This energy is obtained from internal heat exchanges of storage unit in discharging and charging time. For different operating conditions, thermal charging and discharging powers of the storage unit are depicted in Fig. 7.45, Fig. 7.46, Fig. 7.40 and Fig. 7.41. A comparison between numerical simulations and experiments shown a very good agreement. Thereby in this section, thermal energy obtained from numerical simulations is compared with experiments.

Table 7.5: Energy comparison between simulations and experiments during charging and discharging of thermal storage unit.

	Charging		Discharging	
Mass flow rates	5 kg/min	10 kg/min	5 kg/min	10 kg/min
Energy in Exp.	10.28 MJ	8.26 MJ	6.84 MJ	5.87 MJ
Energy in Sim.	9.70 MJ	8.56 MJ	6.93 MJ	6.05 MJ
Difference	5 %	3.5 %	1.3 %	3 %

The comparison between computed energy in numerical simulations and experiments have shown a very good acceptance. LTNE model predicts charging and discharging of the present storage unit within an error estimate of 5 %. This minimal error is resultant of many simplified assumptions in the numerical simulations. Therefore, the numerical model to compute charging and discharging of a macro-encapsulated LHTES unit is employed to optimize the thermal performance.

8 Parametric study for an efficient thermal storage configuration

Thermal efficiency of a LHTES unit can greatly be improved by encapsulation of PCM. But, encapsulation type and storage configuration also play an important role here. Therefore, in the present section of work different storage configurations are studied with the LTNE model. For this, the influence of capsule wall thickness and capsule material is studied for the storage shown in Fig. 6.16 and Fig. 6.14.

Capsule wall thickness

To study thermal response of LHTES unit, a variational study with thinner capsule wall is performed. In these computations, the influence of capsule thickness 1 mm is investigated. Due to thinner capsule, more PCM can be accommodated in PCM capsules with 40 mm external diameter and 250 mm length. Compared to the storage unit with capsule thickness 3 mm, approximately 5 kgs of $MgCl_2 \cdot 6H_2O$ can be additionally accommodated through wall thickness 1 mm. Alongside more PCM mass, a lower capsule mass is implemented in the numerical simulations with capsule thickness of 1 mm. For same operating conditions as in Fig. 7.37, thermal charging and discharging of the storage unit with capsule thickness 1 mm is compared with storage unit with capsule thickness 3 mm. Different mass flow rates i.e. 5 kg/min and 10 kg/min are employed to illustrate the influence of different capsule configuration. Due to the low thermal resistance from capsule wall, an enhanced heat transfer is noted for capsule with thickness 1 mm. But, from an increase in the PCM mass by 25 % delayed charging and discharging is observed. The delay in discharging time of the storage unit is 20 min for an operating mass flow rate of 5 kg/min as in Fig. 8.1 and 25 min for mass flow rate 10 kg/min. Similarly, the delay in charging time of the storage unit is 9 min for an operating mass flow rate of 5 kg/min as seen in Fig. 8.1 and 13 min for mass flow rate 10 kg/min. The comparison of mean operating power and energy is shown in Tab. 8.1.

Table 8.1: Mean thermal power of storage unit with different capsule wall thickness.

	Charging		Discharging	
Mass flow rates	5 kg/min	10 kg/min	5 kg/min	10 kg/min
Mean power 1 mm	1.185 kW	1.365 kW	0.873 kW	1.119 kW
Energy 1 mm	9.99 MJ	10.56 MJ	7.89 MJ	7.67 MJ
Mean power 3 mm	1.025 kW	1.200 kW	0.859 kW	1.115 kW
Energy 3 mm	9.70 MJ	8.54 MJ	6.93 MJ	6.05 MJ

The specific storage density of these different storage configurations vary based on the PCM mass inside a storage unit. Following Pinnau [214], the storage density of the present LHTES unit with cylindrical capsules having wall thickness of 3 mm is 30 kWh/m^3 and approximately 37 kWh/m^3 for the storage with capsules having a wall thickness 1 mm.

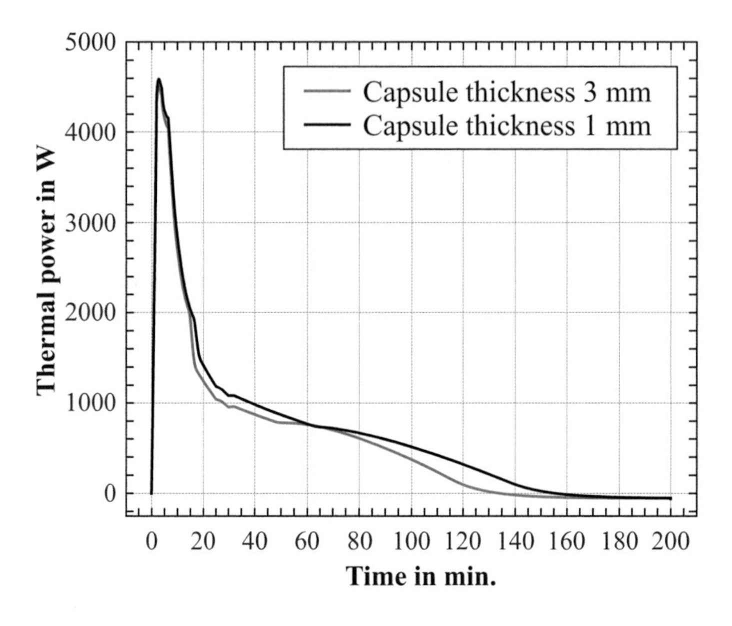

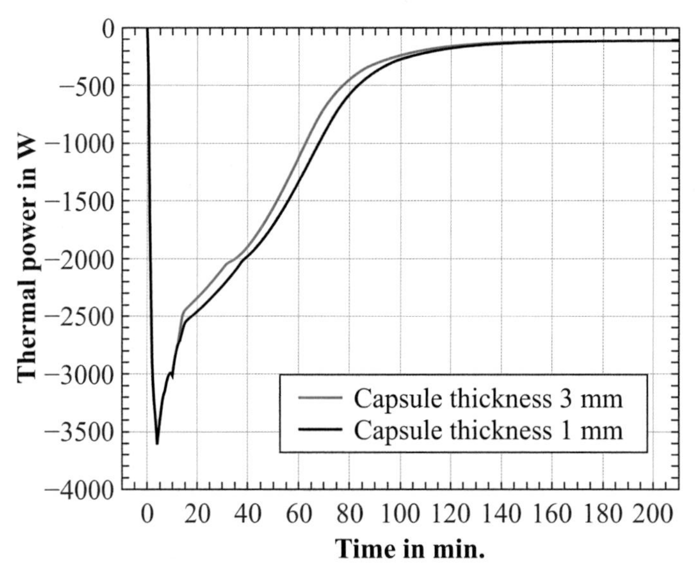

Figure 8.1: Thermal power comparison of storage unit with different capsule wall thickness, on top the discharging power and on bottom the charging power for mass flow rate 5 kg/min.

Thermal resistance offered by capsule wall is reduced by 50 % with 1 mm thick capsules compared to capsules with wall thickness 3 mm. This lower resistance from the capsule wall results in the rise of mean overall heat transfer coefficient value by 4 % for charging and 3 % for discharging process. Nevertheless, from an increase of PCM mass, charging and discharging is delayed for storage unit with capsule thickness 1 mm.

Capsule material

Aluminium capsules as shown in Fig. 6.15 were employed to store PCM in the laboratory scale LHTES unit. But, due to their cost inefficiency employing such aluminium capsules is not economical. Therefore, the possibility to employ other capsule materials is investigated here. For this purpose, standard construction steel is employed as a reference material.

Identical operating conditions as in Fig. 7.37 are employed to compute charging and discharging with new storage configuration. The PCM capsule has an external diameter of 40 mm, length of 250 mm and a wall thickness of 3 mm. Simulation results with construction steel as capsule wall are compared with the results from aluminium. The difference in charging and discharging of thermal storage unit from different capsule materials is shown in Fig. 8.2. Here, it is noticed that the charging and discharging of thermal storage unit is delayed with construction steel as capsule material. It is known that construction steel has a thermal conductivity value of 48 W/mK, which is 4.375 times lesser than thermal conductivity of aluminium ranging at 210 W/mK. Therefore, the resistance offered by the capsules made of construction steel is also approximately 4.375 times larger than that of aluminium. Here, larger thermal resistance from construction steel deteriorates the heat transfer between PCM capsules and HTF, which delays the charging and discharging process. The mean charging and discharging powers for different storage configurations are depicted in Tab. 8.2.

Table 8.2: Mean thermal power of storage unit with different capsule materials.

	Charging		Discharging	
Mass flow rates	5 kg/min	10 kg/min	5 kg/min	10 kg/min
Mean power construction steel	0.986 kW	1.176 kW	0.849 kW	1.103 kW
Energy construction steel	10.13 MJ	8.96 MJ	7.01 MJ	6.42 MJ
Mean power aluminium	1.025 kW	1.200 kW	0.859 kW	1.115 kW
Energy aluminium	9.70 MJ	8.54 MJ	6.93 MJ	6.05 MJ

The mean charging and discharging powers are faded up to 2 % by with construction steel as capsule material. Additionally, 5 % of energy is stored in the storage unit from construction steel compared to aluminium. This is due to the higher mass of steel capsules from their larger density. The full discharging process with an operating mass flow rate of 5 kg/min is delayed by time t = 10 min compared to aluminium. The same is delayed by time t = 7 min for an operating mass flow rate 10 kg/min.

Here, the full charging process with an operating mass flow rate of 5 kg/min is delayed by time t = 11 min with construction steel compared to aluminium. Similarly, the full charging process with an operating mass flow rate of 10 kg/min is delayed by time t = 9 min. As seen in Fig. 8.2, capsule material has a very minimal influence on the charging and discharging power. Therefore, a coupled thermo-economic evaluation would be necessary to evaluate the effective way to optimize a macro-encapsulated LHTES unit.

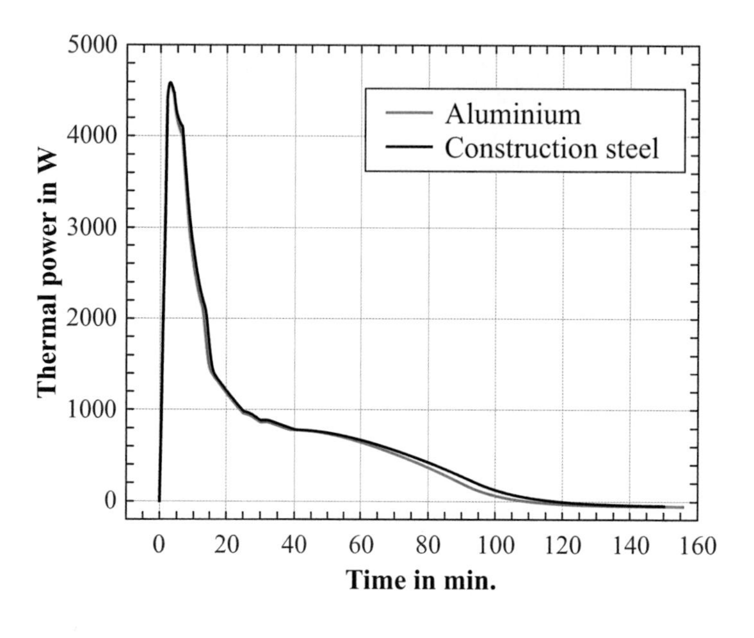

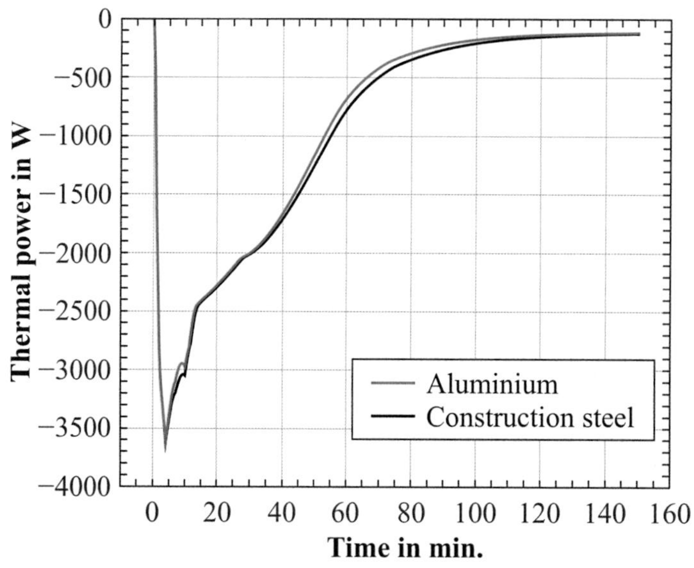

Figure 8.2: Thermal power comparison of storage unit with different capsule materials, on top the discharging power and on bottom the charging power for mass flow rate 5 kg/min.

9 Summary

To meet the energy demands of future, energy storage and conversion are essential steps. Storing thermal energy for future use, does not only increase the system efficiency but also supports global warming reduction. In this scenario, latent heat thermal energy storage (LHTES) units are very relevant in domestic, industrial and mobile applications. Phase change materials (PCM) with low thermal conductivity result in a poor charging and discharging of the LHTES units. This low charging and discharging power can be improved by macro-encapsulation of PCM. But, the benefit of macro-encapsulation depends on PCM capsule size and geometrical configuration. Therefore, the present work employs numerical modelling to study of different parameters of LHTES unit.

Numerical modelling and simulation enables to investigate the response of macro-encapsulated LHTES units with different storage parameters. Thereby, in the present work different numerical models are developed to represent LHTES unit at multiple length scales. Accordingly, a numerical model for small scale PCM capsule predicts the detailed solid-liquid phase change, whereas the flow of heat transfer fluid (HTF) in storage unit including heat transfer is modelled with a simplified numerical model. A detailed heat transfer model is developed to represent the heat transfer and fluid flow in a container for a row of capsules.

In this work, the multiphase numerical model predicts PCM melting in a small scale cubical cavity very well. Here, the dimensions of solid PCM in numerical simulations also looks similar to experiments. From the numerical results, it is observed that air cavity in PCM capsule suppresses the heat flow from top wall to PCM. On the other side, close contact melting accelerates the melting process by contributing a large heat flow into the cavity. The increase of pressure inside the cavity as seen in numerical simulations is not observed in experiments. In experiments, this behaviour is influenced from the air outflow through pressure vents. Due to its large computational time, the multiphase numerical model for a small scale PCM capsule is simplified. Here, two different simplified models are developed to represent the melting of PCM in a cubical cavity. The simplified convective model reduces the multiphase PCM-Air system into single phase model, by implementing air as a convective boundary condition. A comparison with experiments show that the simplified convective model also represents the melting of PCM very well. Whereas, the simplified conductive model represents the close contact melting and natural convection in PCM with an effective thermal conductivity correlation. For this correlation, constants c and n are adapted for each set-up. Despite this strong simplification, the simplified conductive model shows a decent agreement with experiments.

Paraffin used in the cubical cavity test case is not suitable, to operate the LHTES unit at high temperatures. Therefore, phase change of a salt-hydrate with a higher melting temperature is investigated using numerical simulations and experiments. Here, the role of convection and close contact melting of salt hydrate in a cylindrical tube is investigated. Initially, only the role of natural convection is estimated using the simplified conductive model with an empirical correlation. The numerical results are compared with experiments, which have shown a very good acceptance. Later, the natural convection along with solid settling is estimated with a fitted correlation for the present geometrical and operating configuration.

For an intermediate scale, the conjugate heat transfer between PCM capsules and HTF is investigated. For this, cylindrical capsules filled with paraffin as PCM are employed. Forced convection induced heat transfer in HTF, PCM phase change with close contact melting including natural convection in capsules and capsule wall conduction is represented with this model. The CHT model predicts the unsteady heat transfer in HTF along with the PCM melting in the capsule very similar to experiments. As in Fig. 6.10, for horizontally placed capsule row a density layered flow is observed at initial instances of the charging process, see Fig. 7.35. By employing the CHT model, a detailed heat transfer inside a LHTES unit can be investigated.

For a large scale macro-encapsulated LHTES unit the LTNE model is very appropriate. Thermal charging and discharging of the present storage unit is represented very well using this numerical model. When compared with experiments, the simulation results from LTNE model have shown a good agreement. Here, it is noticed that the forced convection in HTF also influence the heat transfer around PCM capsules. For the laboratory scale LHTES unit with vertically positioned cylindrical capsules, heat transfer coefficient between HTF and capsule wall increases with the increment in mass flow rate. Therefore, a sooner charging and discharging is observed for higher operating mass flow rates. For vertically placed capsule columns, HTF flow direction also process play an important role in charging and discharging as seen in Fig. 6.17. For charging process, heat transfer inside the PCM capsule is implemented with a validated effective thermal conductivity correlation for a cylindrical enclosure. The effective thermal conductivity in this case is an approximate of the close contact melting and natural convection role in the geometry. Therefore, some deviation between experiments and simulation is noted. Due to its strong simplification, such deviations can be permitted for such a large storage unit. Nevertheless, the discharging process of the storage unit is very well represented with numerical simulations as in experiments. With the help of validated numerical models, the response of LHTES unit with different capsule parameters was investigated.

Key questions posed in the framework of this work are answered as follows:

1. How evolved is the field of numerical simulation to substantially study the behaviour of LHTES units at different length scales?

In the present day scenario, numerical modelling and simulation offers a wide range of application fields along with strong system optimization potential. In the present work, multiscale thermo-fluidic modelling of a macro-encapsulated LHTES unit is illustrated. Here, numerical modelling and simulation is employed to represent physical phenomena at different length scales. Validation of these numerical models illustrate the potential of modelling and simulation to study physical processes in a storage unit at multiple scales. The present work illustrates the ability of numerical modelling to represent the detailed melting of PCM in a small scale capsule alongside the thermal charging and discharging of a large scale LHTES unit. These validated numerical models can be employed to optimize the geometrical configuration of PCM capsule alongside the storage configuration of a large scale LHTES unit.

2. How can physical process in a small scale PCM capsule be represented in a large scale macro-encapsulated LHTES unit?

In this work, solid-liquid phase change in a small scale PCM capsule is expressed with an effective thermal conductivity in a large scale macro-encapsulated LHTES unit. The effective thermal conductivity is evaluated by validating the phase change of PCM in a small capsule with experiments. During the charging of a storage unit, melting process influenced by natural convection and close contact melting is represented here. For discharging process, solidification of PCM inside the capsule is represented with a simple thermal conductivity of solid PCM. The capsule packed bed in a LHTES unit is modelled by volume averaged approach. The heat transfer between each PCM capsule and HTF is modelled with volumetric overall heat transfer coefficient, which is computed from thermal resistances. Thermal resistance in PCM is computed from the phase change fraction in PCM.

3. What is the role of natural convection and close contact melting in a macro-capsule?

In a melting process, liquid natural convection enhances the heat transfer inside a PCM capsule. Heat transfer in a melting process can be further accelerated, when solid PCM settles to the capsule bottom. But, the increase in PCM melting rate from solid settling purely depends on the bottom wall temperature and its surface area. For a cubical cavity the fraction of heat flow through bottom wall may rise from 25 % to 95 % of total heat flux as shown in Fig. 7.14. Such a behaviour is observed when the bottom wall surface area contributes to a countable amount of total surface area. For a long cylindrical enclosure in Fig. 7.26 and Fig. 7.27, solid settling results in a slight decrease of melting rate than with no solid settling. For such long cylinder, the heat transfer surface of bottom wall contributes approximately 4 % of the total cylinder surface area. Here, due to solid settling the solid PCM gives up the heat flow from side wall for the heat flow from bottom wall. Therefore, close contact melting can only be exploited for geometries with large bottom surface area to increase the melting rate.

4. Which factors need to be taken into account to develop an efficient macro-encapsulated LHTES unit?

It is observed that higher mass flow rates result in a sooner charging and discharing of a LTNE storage unit. To develop a techno-economic storage unit, inexpensive capsule material with low wall thickness is an effective solution. Geometrical configuration of the PCM capsule also plays a vital role, due to its strong influence on external heat transfer. A very thin air cavity in PCM capsule will be beneficial for an effective charging and discharging. Because, the air cavity obstructs heat flow into the PCM capsule. For a storage with horizontally placed capsules, higher mass flow rates solve the density layering at initial instances.

Deviations between the numerical simulations and experiments can be investigated for the further research. Importantly, the CHT model developed in this work assists to understand the detailed heat transfer behaviour in the storage unit, but the model needs a large computational time. This could be further optimized for a low computational time and efficiency. The numerical simulations with LTNE model also show a slight deviation from experiments for charging process. But, the discharging process with the same model can be represented very well similar to experiments. This deviation in the numerical results of charging process originates from the effective thermal conductivity correlation, which takes an ideal melting process into account. But in reality an ideal melting process with proper close contact melting is not often observed. For further understanding, this influence of close contact melting relating with effective thermal conductivity can be investigated in detail.

10 Zusammenfassung

Die Energieumwandlung und -speicherung sind wichtige Schritte um den zukünftigen Energiebedarf zu decken. Die Zwischenspeicherung thermischer Energie verbessert nicht nur die Systemeffizienz, sondern hilft auch bei der Reduzierung der globalen Erwärmung. Hierbei können latente thermische Speicher (LTS) einen wichtigen Beitrag in häuslichen, industriellen und mobilen Anwendungen leisten. Die geringe Wärmeleitfähigkeit von Phasenwechselmaterialien (PCM) führt jedoch zu geringen Be- und Entladeleistungen des LTS und kann durch Makroverkapselung verbessert werden. Dieser Vorteil der Makroverkapselung ist von der Größe und der geometrischen Anordnung der Kapseln abhängig. Die vorliegende Arbeit untersucht daher mithilfe von numerischen Simulationen den Einfluss verschiedener Parameter auf den LTS.

Die Modellierung und Simulation ermöglicht die Untersuchung von unterschiedlichen Speicherparametern auf das Verhalten des makroverkapselten LTS. Hierfür werden in der vorliegenden Arbeit verschiedene numerische Modelle entwickelt, die den LTS in unterschiedlichen Größenskalen beschreiben. Ein numerisches Modell für kleinskalige PCM-Kapseln beschreibt den detaillierten fest/flüssig Phasenwechsel. Demgegenüber wird die Wärmeübertragung durch das Wärmeträgerfluid (WTF) in einem kompletten Speicher mit Vereinfachungen modelliert. Ein detailliertes Modell kommt zur Beschreibung der Wärmeübertragung in einen Behälter mit einzelnen, in Reihe angeordneten Kapseln zum Einsatz.

Das in dieser Arbeit verwendete Mehrphasenmodell liefert eine sehr gute Beschreibung des Aufschmelzvorganges in einer Würfelgeometrie. Die Form des Feststoffes stimmen in Simulationen und Experimenten gut überein. Die numerischen Ergebnisse zeigen, dass die Luftschicht in der PCM-Kapsel den Wärmestrom von der oberen Wand zum PCM hemmt. Demgegenüber verursacht das Kontaktschmelzen einen großen Wärmestrom und führt so zu einem beschleunigten Aufschmelzprozess. Ein Druckanstieg innerhalb der Kapseln, wie er bei den Simulationen zu Sehen ist, konnte in den Experimenten nicht beobachtet werden, da die hier vorliegenden Luftauslässe den Innendruck beeinflussen. Da Mehrphasenmodell für PCM-Kapseln wurde aufgrund der langen Rechenzeiten vereinfacht. Zwei vereinfachte Modelle wurden entwickelt, um das Aufschmelzen in der Würfelgeometrie darzustellen. Das vereinfachte Konvektionsmodell reduziert das Mehrphasen-PCM-Luft-System in ein Einphasenmodell, indem die Luft als konvektive Randbedingung implementiert wird. Der Vergleich mit Experimenten zeigt, dass auch mit dem vereinfachten Konvektionsmodell der Aufschmelzvorgang sehr gut wiedergegeben werden kann. Das vereinfachte Wärmeleitfähigkeitsmodell beschreibt das Kontaktschmelzen und die natürliche Konvektion im PCM mit einer Korrelation für die effektive Wärmeleitfähigkeit. Die für die Korrelation benötigten Konstanten wurden für jeden Aufbau ermittelt. Trotz dieser starken Vereinfachungen zeigt auch das vereinfachte Wärmeleitfähigkeitsmodell eine gute Übereinstimmung mit den Experimenten.

Das in der Würfelgeometrie eingesetzte Paraffin eignet sich nicht zum Einsatz in Hochtemperatur-LTS. Daher werden zusätzlich mit einem Salzhydrat Experimente bei höherer Temperatur durchgeführt und mit numerische Simulationen verglichen. Der Einfluss der Konvektion und des Kontaktschmelzens beim Salzhydrat wurde hierbei in einem zylindrischen Rohr untersucht. Zunächst wurde nur der Einfluss der natürlichen Konvektion mit dem vereinfachten Wärmeleitfähigkeitsmodell und einer empirischen Korrelation ermittelt. Der Vergleich der numerischen Ergebnisse mit Experimenten zeigte dabei eine sehr gute Übereinstimmung. Im Anschluss daran wurde die natürliche Konvektion zusammen mit dem Absinken des Feststoffes durch eine Korrelation für die vorliegende Geometrie und die Betriebsbedingungen beschrieben.

Für die mittlere Skala wird die Wärmeübertragung zwischen den Speicherkapseln und dem WTF untersucht. Hierfür kommen zylindrische Kapseln, welche mit einem Paraffin als PCM gefüllt sind, zum Einsatz. Das Modell beschreibt die Wärmeübertragung durch die erzwungene Konvektion des WTF, den Phasenwechsel mit Kontaktschmelzen und natürlicher Konvektion in den Kapseln sowie die Wärmeleitung in den Kapselwänden. Das CHT-Modell prognostiziert die instationäre Wärmeübertragung im WTF und das Aufschmelzen des PCM in den Kapseln in sehr ähnlicher Weise, wie es auch die Experimente zeigen. In Fig. 6.10 ist zu erkennen, dass sich bei der horizontal angeordneten Zylinderreihe zu Beginn des Beladevorganges eine Schichtströmung ausbildet (Fig. 7.35). Mit dem CHT-Modell kann die Wärmeübertragung innerhalb eines LTS sehr detailliert untersucht werden.

Für großskalige makroverkapselte LTS ist das LTNE-Modell geeignet. Die thermische Be- und Entladung des vorgestellten Speichers kann mit diesem numerischen Modell sehr gut dargestellt werden. Der Vergleich mit Experimenten zeigt eine gute Übereinstimmung mit den Ergebnissen des LTNE-Modells. Es kann festgestellt werden, dass die erzwungene Konvektion des WTF die Wärmeübertragung an den Kapseln beeinflusst. Im Laborspeicher mit vertikal positionierten, zylindrischen Kapseln erhöht sich der Wärmeübergangskoeffizient zwischen WTF und Kapselwand mit zunehmendem Massenstrom. Daher ist die Be- und Entladung beim Betrieb mit größeren Massenströmen schneller abgeschlossen. Weiterhin hat die Durchströmungsrichtung durch die vertikal angeordneten Kapselreihen einen großen Einfluss auf den Be- und Entladevorgang (Fig. 6.17). Für das Beladen ist die Wärmeübertragung innerhalb der Kapsel mit einer validierten Korrelation für die effektive Wärmeleitfähigkeit in zylindrischen Behältern implementiert. Die effektive Wärmeleitfähigkeit beschreibt in diesem Fall eine Annäherung an das Kontaktschmelzen und die natürliche Konvektion in der untersuchten Geometrie. Daher sind auch Abweichungen zwischen den Experimenten und den Simulationen festzustellen. Aufgrund der starken Vereinfachungen sind solche Abweichungen für große Speicher zulässig. Trotzdem stimmen die Simulationen des Entladevorganges sehr gut mit den Ergebnissen aus den Experimenten überein. Der Entladevorgang des Speichers wird sehr gut durch die Simulationen wie auch in den Experimenten beschrieben. Mit Hilfe der validierten numerischen Modelle wird abschließend das Verhalten des makroverkapselten LTS durch Parameterstudien zur Verkapselung untersucht.

Die Antworten auf die in dieser Arbeit gestellten Schlüsselfragen sind im Folgenden zu finden:

1. Wie gut ist das Arbeitsgebiet der numerischen Simulation entwickelt, um das grundlegende Verhalten von LTS unterschiedlicher Größenordnung zu untersuchen?

Die numerische Simulation kann gegenwärtig für einen breiten Anwendungsbereich eingesetzt werden und bietet ein großes Potential zur Systemoptimierung. In der vorliegenden Arbeit wird die thermofluiddynamische Modellierung von makroverkapselten LTS unterschiedlicher Größenordnung vorgestellt. Die numerische Modellierung und Simulation wird eingesetzt, um physikalische Effekte in verschiedenen Größenordnungen darzustellen. Die Validierung dieser Modelle verdeutlicht das Potential der Modellierung und Simulation zum Verständnis der physikalischen Vorgänge in Speichern unterschiedlicher Größe. Die Arbeit zeigt die Möglichkeiten der numerischen Modellierung, von detaillierten Aufschmelzprozessen in einzelnen Kapseln bis hin zu der Untersuchung des Be- und Entladeverhaltens ganzer LTS. Die validierten Modelle können zur Optimierung von Kapselgeometrien und deren Anordnung in großen LTS eingesetzt werden.

2. Wie können die in kleinen Kapseln ablaufenden, physikalischen Vorgänge auf große makroverkapselte LTS übertragen werden?

In dieser Arbeit wird der fest-flüssig Phasenwechsel einzelner Speicherkapseln mithilfe einer effektiven Wärmeleitfähigkeit im großen, makroverkapselten LTS implementiert. Die effektive Wärmeleitfähigkeit wird anhand von Validierungsexperimenten zum Phasenwechsel in kleinen Kapseln bestimmt. Das beim Schmelzprozess auftretende Kontaktschmelzen und die natürliche Konvektion werden hierbei berücksichtigt. Für den Entladevorgang wird die Erstarrung des PCM innerhalb der Kapsel mit der Wärmeleitfähigkeit des Feststoffs beschrieben. Die Kapselpackung in einem LTS wird durch einen volumengemittelten Ansatz modelliert. Die Wärmeübertragung zwischen den Kapseln und dem WTF ist durch einen volumetrischen Wärmedurchgangskoeffizienten modelliert, welcher aus den thermischen Widerständen berechnet wird. Der thermische Widerstand des PCM wird aus dem Phasenanteil an umgewandelten PCM ermittelt.

3. Welchen Einfluss haben die natürliche Konvektion und das Kontaktschmelzen in Makrokapseln?

Beim Aufschmelzen erhöht die natürliche Konvektion die Wärmeübertragung in der PCM-Kapsel. Die Wärmeübertragung kann weiterhin durch das Absinken der Feststoffphase auf den Kapselboden verbessert werden, wobei die Zunahme der Schmelzrate hierbei von der Temperatur des Kapselbodens und dessen Oberfläche abhängt. Für die Würfelgeometrie kann der Anteil des Wärmestromes durch die untere Wand, wie in Fig. 7.14 gezeigt von anfänglich 25 % bis auf 95 % des gesamten Wärmestromes ansteigen. Dieses Verhalten kann beobachtet werden, wenn die Oberfläche der unteren Wand einen signifikanten Anteil der gesamten Oberfläche einnimmt. Für lange, zylindrische Kapseln, wie sie beispielsweise in Fig. 7.26 und Fig. 7.27 gezeigt sind, führt das Absinken verglichen mit dem Fall ohne Absinken hingegen zu einem leichten Rückgang der Schmelzrate. Bei diesen langen Zylindern beträgt der Anteil der unteren Wand nur etwa 4 % der gesamten Zylinderoberfläche. In diesem Fall führt das Absinken zu einem reduzierten Wärmestrom durch die seitlichen Wände. Kontaktschmelzen kann also nur bei Geometrien mit ausreichend großen Bodenflächen genutzt werden, um die Schmelzrate zu steigern.

4. Welche Einflussgrößen müssen für die Entwicklung effizienter makroverkapselter LTS berücksichtigt werden?

Es zeigt sich, dass größere Massenströme zur schnelleren Be- und Entladung eines LTS führen. Zur Entwicklung eines techno-ökonomisch sinnvollen Speichers werden günstige Kapselwerkstoffe mit geringen Wandstärken benötigt. Die geometrische Anordnung der Kapseln spielt eine große Rolle für die externe Wärmeübertragung. Ein möglichst kleines Luftpolster in der Kapsel bewirkt eine effektive Be- und Entladung, da die Luftschicht den Wärmestrom in die Kapsel sonst behindert. Für Speicher mit horizontal eingesetzten Kapseln können höhere Massenströme die zu Beginn vorliegende Dichte-Schichtung beheben.

Abweichungen zwischen den numerischen Simulationen und den Experimenten sollten in Folgearbeiten untersucht werden. Das in der Arbeit entwickelte CHT-Modell liefert einen wichtigen Beitrag zum Verständnis der in Speichern ablaufenden Wärmeübertragungsvorgänge. Das Modell benötigt zur Simulation des Be- und Entladeverhaltens bisher allerdings lange Rechenzeiten und sollte deshalb hinsichtlich Rechenzeit und -effizienz noch weiter optimiert werden. Die Simulationen mit dem LTNE-Modell zeigen beim Beladen noch geringe Abweichungen zu den Experimenten, wohingegen der Entladevorgang mit dem gleichen Modell mit sehr guter Übereinstimmung wiedergegeben werden kann. Diese Abweichungen der numerischen Ergebnisse beim Beladevorgang werden durch die verwendete Korrelation für die effektive Wärmeleitfähigkeit verursacht, welche einen idealisierten Schmelzprozess voraussetzt. In der Realität kann dieser idealisierte Fall mit idealem Kontaktschmelzen allerdings nicht beobachtet werden. Zum besseren Verständnis sollte der Einfluss des Kontaktschmelzens auf die effektive Wärmeleitfähigkeit tiefergehend untersucht werden.

Bibliography

[1] T Jayaraman and Tejal Kanitkar. The paris agreement. *Economic and Political Weekly*, 51(3), 2016.

[2] Rana Adib, HE Murdock, F Appavou, A Brown, B Epp, A Leidreiter, C Lins, HE Murdock, E Musolino, K Petrichenko, et al. Renewables 2016 global status report. *Global status report renewable energy policy network for the 21st century (REN21)*, 2016.

[3] Nabeel S Dhaidan and JM Khodadadi. Melting and convection of phase change materials in different shape containers: A review. *Renewable and Sustainable Energy Reviews*, 43:449–477, 2015.

[4] A. Felix Regin, S.C. Solanki, and J.S. Saini. Heat transfer characteristics of thermal energy storage system using pcm capsules: A review. *Renewable and Sustainable Energy Reviews*, 12(9):2438 – 2458, 2008.

[5] Dagmar Oertel. Energiespeicher–stand und perspektiven. *Sachstandsbericht zum Monitoring „Nachhaltige Energieversorgung ", Arbeitsbericht*, (123), 2008.

[6] Randall D Knight. *Physics for scientists and engineers*, volume 1. Pearson, 2004.

[7] J. Anderson. *Computational Fluid Dynamics*. McGraw-Hill Education, 1995.

[8] Joel H. Ferziger and Milovan Peric. *Computational Methods for Fluid Dynamics*, volume 3. Springer-Verlag Berlin Heidelberg, 2002.

[9] Ruediger Schwarze. *CFD-Modellierung: Grundlagen und Anwendungen bei Stroemungsprozessen*. Springer-Verlag, 2012. (in German).

[10] Fridtjov Irgens. *Continuum mechanics*. Springer Science & Business Media, 2008.

[11] Joseph Boussinesq. *Théorie de l'écoulmnent tourbillonnant et tumultuex des liquides dans les lits rectilignes à grande section*, volume 1. Gauthier-Villars, 1897.

[12] Junuthula Narasimha Reddy. *An introduction to continuum mechanics*. Cambridge University Press, 2013.

[13] John Crank and Phyllis Nicolson. A practical method for numerical evaluation of solutions of partial differential equations of the heat-conduction type. *Advances in Computational Mathematics*, 6(1):207–226, 1996.

[14] A.J.M. Spencer. *Continuum Mechanics*. Dover Publications, 2012.

[15] Richard Courant, Eugene Isaacson, and Mina Rees. On the solution of nonlinear hyperbolic differential equations by finite differences. *Communications on Pure and Applied Mathematics*, 5(3):243–255, 1952.

[16] P.A. Nikrityuk. *Computational Thermo-Fluid Dynamics: In Materials Science and Engineering*. Wiley, 2011.

[17] Suhas V Patankar. Numerical heat transfer and fluid flow. *Computational methods in mechanics and thermal science*, page 197, 1980.

[18] Richard Courant, Kurt Friedrichs, and Hans Lewy. Über die partiellen differenzengleichungen der mathematischen physik. *Mathematische annalen*, 100(1):32–74, 1928.

[19] Christopher J. Greenshields. Openfoam programmers guide. Online, OpenFOAM Foundation Ltd., 2011-2015.

[20] J Stefan. Über die theorie der eisbildung. *Monatshefte für Mathematik*, 1(1):1–6, 1890.

[21] PV Danckwerts. Unsteady-state diffusion or heat-conduction with moving boundary. *Transactions of the Faraday Society*, 46:701–712, 1950.

[22] J Crank. Two methods for the numerical solution of moving-boundary problems in diffusion and heat flow. *The Quarterly Journal of Mechanics and Applied Mathematics*, 10(2):220–231, 1957.

[23] RM Furzeland. A comparative study of numerical methods for moving boundary problems. *IMA Journal of Applied Mathematics*, 26(4):411–429, 1980.

[24] AB Crowley. On the weak solution of moving boundary problems. *IMA Journal of Applied Mathematics*, 24(1):43–57, 1979.

[25] Comini Bonacina, G Comini, A Fasano, and M 1973 Primicerio. Numerical solution of phase-change problems. *International Journal of Heat and Mass Transfer*, 16(10):1825–1832, 1973.

[26] G Comini, S Del Guidice, RW Lewis, and OC Zienkiewicz. Finite element solution of non-linear heat conduction problems with special reference to phase change. *International Journal for Numerical Methods in Engineering*, 8(3):613–624, 1974.

[27] K Morgan, RW Lewis, and OC Zienkiewicz. An improved algrorithm for heat conduction problems with phase change. *International Journal for Numerical Methods in Engineering*, 12(7):1191–1195, 1978.

[28] N Shamsundar and EM Sparrow. Analysis of muidimensional conduction phase change via the enthalpy model. *Journal of Heat Transfer*, page 333, 1975.

[29] D.K.Gartling. Finite element analysis of convective heat transfer problems with change of phase. In *Proceedings of the Conference on numerical methods in laminar and turbulent flows, Swansea, UK*, 1978.

[30] K. Morgan. A numerical analysis of freezing and melting with convection. *Computer Methods in Applied Mechanics and Engineering*, 28(3):275 – 284, 1981.

[31] V Voller and M Cross. Accurate solutions of moving boundary problems using the enthalpy method. *International Journal of Heat and Mass Transfer*, 24(3):545–556, 1981.

[32] W Donald Rolph and Klaus-JÜRgen Bathe. An efficient algorithm for analysis of nonlinear heat transfer with phase changes. *International Journal for Numerical Methods in Engineering*, 18(1):119–134, 1982.

[33] Karl-Hermann Tacke. Discretization of the explicIt enthalpy method for planar phase change. *International journal for numerical methods in engineering*, 21(3):543–554, 1985.

[34] J. Crank. *Free and Moving Boundary Problems*. Oxford science publications. Clarendon Press, 1987.

[35] QT Pham. The use of lumped capacitance in the finite-element solution of heat conduction problems with phase change. *International Journal of Heat and Mass Transfer*, 29(2):285–291, 1986.

[36] VR Voller, M Cross, and NC Markatos. An enthalpy method for convection/diffusion phase change. *International journal for numerical methods in engineering*, 24(1):271–284, 1987.

[37] Vaughan R Voller and C Prakash. A fixed grid numerical modelling methodology for convection-diffusion mushy region phase-change problems. *International Journal of Heat and Mass Transfer*, 30(8):1709–1719, 1987.

[38] C Beckermann and R Viskanta. Natural convection solid/liquid phase change in porous media. *International journal of heat and mass transfer*, 31(1):35–46, 1988.

[39] AD Brent, VR Voller, and KTJ Reid. Enthalpy-porosity technique for modeling convection-diffusion phase change: application to the melting of a pure metal. *Numerical Heat Transfer, Part A Applications*, 13(3):297–318, 1988.

[40] C Gau and R Viskanta. Melting and solidification of a metal system in a rectangular cavity. *International Journal of Heat and Mass Transfer*, 27(1):113–123, 1984.

[41] WD Bennon and FP Incropera. Numerical analysis of binary solid-liquid phase change using a continuum model. *Numerical Heat Transfer, Part A Applications*, 13(3):277–296, 1988.

[42] RW Lewis and PM Roberts. Finite element simulation of solidification problems. *Applied Scientific Research*, 44(1-2):61–92, 1987.

[43] Jonathan A Dantzig. Modelling liquid–solid phase changes with melt convection. *International Journal for Numerical Methods in Engineering*, 28(8):1769–1785, 1989.

[44] Yiding Cao, Amir Faghri, and Won Soon Chang. A numerical analysis of stefan problems for generalized multi-dimensional phase-change structures using the enthalpy transforming model. *International journal of heat and mass transfer*, 32(7):1289–1298, 1989.

[45] Y Cao and A Faghri. A numerical analysis of phase-change problems including natural convection. *Journal of Heat Transfer*, 112(3):812–816, 1990.

[46] M Okada. Analysis of heat transfer during melting from a vertical wall. *International journal of heat and mass transfer*, 27(11):2057–2066, 1984.

[47] M Lacroix and VR Voller. Finite difference solutions of solidification phase change problems: transformed versus fixed grids. *Numerical Heat Transfer*, 17(1):25–41, 1990.

[48] VR Voller, AD Brent, and C Prakash. Modelling the mushy region in a binary alloy. *Applied mathematical modelling*, 14(6):320–326, 1990.

[49] C Prakash and V Voller. On the numerical solution of continuum mixture model equations describing binary solid-liquid phase change. *Numerical Heat Transfer, Part B Fundamentals*, 15(2):171–189, 1989.

[50] CR Swaminathan and VR Voller. A general enthalpy method for modeling solidification processes. *Metallurgical transactions B*, 23(5):651–664, 1992.

[51] CR Swaminathan and VR Voller. On the enthalpy method. *International Journal of Numerical Methods for Heat & Fluid Flow*, 3(3):233–244, 1993.

[52] VR Voller and CR Swaminathan. Treatment of discontinuous thermal conductivity in control-volume solutions of phase-change problems. *Numerical Heat Transfer, Part B Fundamentals*, 24(2):161–180, 1993.

[53] R Viswanath and Y Jaluria. A comparison of different solution methodologies for melting and solidification problems in enclosures. *Numerical Heat Transfer, Part B Fundamentals*, 24(1):77–105, 1993.

[54] Zhen-Xiang Gong and Arun S Mujumdar. Noniterative procedure for the finite-element solution of the enthalpy model for phase-change heat conduction problems. *Numerical Heat Transfer*, 27(4):437–446, 1995.

[55] QT Pham. Comparison of general-purpose finite-element methods for the stefan problem. *Numerical Heat Transfer*, 27(4):417–435, 1995.

[56] W Minkowycz. *Advances in numerical heat transfer*, volume 1. CRC press, 1996.

[57] W. Shyy, H.S. Udaykumar, M.M. Rao, and R.W. Smith. *Computational Fluid Dynamics with Moving Boundaries*. Dover Books on Engineering. Dover Publications, 2007.

[58] DW Morecroft. Reactions of octadecane and decoic acid with clean iron surfaces. *Wear*, 18(4):333–339, 1971.

[59] M. Bareiss and H. Beer. An analytical solution of the heat transfer process during melting of an unfixed solid phase change material inside a horizontal tube. *International Journal of Heat and Mass Transfer*, 27(5):739 – 746, 1984.

[60] MK Moallemi, BW Webb, and R Viskanta. An experimental and analytical study of close-contact melting. *J. Heat Transfer*, 108(4):894–899, 1986.

[61] MK Moallemi and R Viskanta. Analysis of close-contact melting heat transfer. *International journal of heat and mass transfer*, 29(6):855–867, 1986.

[62] EM Sparrow and GT Geiger. Melting in a horizontal tube with the solid either constrained or free to fall under gravity. *International journal of heat and mass transfer*, 29(7):1007–1019, 1986.

[63] Adrian Bejan. Single correlation for theoretical contact melting results in various geometries. *International communications in heat and mass transfer*, 19(4):473–483, 1992.

[64] Adrian Bejan. Contact melting heat transfer and lubrication. *Advances in heat transfer*, 24:1–38, 1994.

[65] Y Asako, M Faghri, M Charmchi, and PA Bahrami. Numerical solution for melting of unfixed rectangular phase-change material under low-gravity environment. *Numerical Heat Transfer*, 25(2):191–208, 1994.

[66] A Shirvanian, M Faghri, Z Zhang, and Y Asako. Numerical solution of the effect of vibration on melting of unfixed rectangular phase-change material under variable-gravity environment. *Numerical Heat Transfer, Part A Applications*, 34(3):257–278, 1998.

[67] M Faghri, Y Asako. Effect of density change on melting of unfixed rectangular phase-change material under low-gravity environment. *Numerical Heat Transfer: Part A: Applications*, 36(8):825–838, 1999.

[68] Y Asako, E Gonçalves, M Faghri, and M Charmchi. Numerical solution of melting processes for fixed and unfixed phase change material in the presence of magnetic field–simulation of low-gravity environment. *Numerical Heat Transfer: Part A: Applications*, 42(6):565–583, 2002.

[69] B Ghasemi and M Molki. Melting of unfixed solids in square cavities. *International journal of heat and fluid flow*, 20(4):446–452, 1999.

[70] Marcel Lacroix. Contact melting of a phase change material inside a heated parallelepedic capsule. *Energy conversion and management*, 42(1):35–47, 2001.

[71] Dominic Groulx and Marcel Lacroix. Effects of convection and inertia on close contact melting. *International journal of thermal sciences*, 42(12):1073–1080, 2003.

[72] Dominic Groulx and Marcel Lacroix. Study of close contact melting of ice from a sliding heated flat plate. *International journal of heat and mass transfer*, 49(23):4407–4416, 2006.

[73] Dominic Groulx and Marcel Lacroix. Study of the effect of convection on close contact melting of high prandtl number substances. *International journal of thermal sciences*, 46(3):213–220, 2007.

[74] E Assis, L Katsman, G Ziskind, and R Letan. Numerical and experimental study of melting in a spherical shell. *International Journal of Heat and Mass Transfer*, 50(9):1790–1804, 2007.

[75] F.L. Tan, S.F. Hosseinizadeh, J.M. Khodadadi, and Liwu Fan. Experimental and computational study of constrained melting of phase change materials (pcm) inside a spherical capsule. *International Journal of Heat and Mass Transfer*, 52(15–16):3464 – 3472, 2009.

[76] H. Shmueli, G. Ziskind, and R. Letan. Melting in a vertical cylindrical tube: Numerical investigation and comparison with experiments. *International Journal of Heat and Mass Transfer*, 53(19):4082 – 4091, 2010.

[77] Fabian Rösler and Dieter Brüggemann. Shell-and-tube type latent heat thermal energy storage: numerical analysis and comparison with experiments. *Heat and mass transfer*, 47(8):1027, 2011.

[78] S.F. Hosseinizadeh, A.A. Rabienataj Darzi, F.L. Tan, and J.M. Khodadadi. Unconstrained melting inside a sphere. *International Journal of Thermal Sciences*, 63:55 – 64, 2013.

[79] Fabian Rösler. *Modellierung und simulation der Phasenwechselvorgänge in makroverkapselten latenten thermischen Speichern*, volume 24. Logos Verlag Berlin GmbH, 2014. (in German).

[80] Yoram Kozak, Tomer Rozenfeld, and Gennady Ziskind. Close-contact melting in vertical annular enclosures with a non-isothermal base: Theoretical modeling and application to thermal storage. *International Journal of Heat and Mass Transfer*, 72:114–127, 2014.

[81] Y Kozak and G Ziskind. Novel enthalpy method for modeling of pcm melting accompanied by sinking of the solid phase. *International Journal of Heat and Mass Transfer*, 112:568–586, 2017.

[82] Raghavendra Rohith Kasibhatla, Andreas König-Haagen, and Dieter Brüggemann. Numerical modelling of wetting phenomena during melting of pcm. *Procedia Engineering*, 157:139 – 147, 2016. Selected Papers from IX International Conference on Computational Heat and Mass Transfer (ICCHMT2016).

[83] Raghavendra Rohith Kasibhatla, Andreas König-Haagen, Fabian Rösler, and Dieter Brüggemann. Numerical modelling of melting and settling of an encapsulated pcm using variable viscosity. *Heat and Mass Transfer*, 53(5):1735–1744, 2017.

[84] SK Friedlander. Behavior of suspended particles in a turbulent fluid. *AIChE Journal*, 3(3):381–385, 1957.

[85] Francis H Harlow and J Eddie Welch. Numerical calculation of time-dependent viscous incompressible flow of fluid with free surface. *The physics of fluids*, 8(12):2182–2189, 1965.

[86] Francis H Harlow, John P Shannon, and J Eddie Welch. Liquid waves by computer. *Science*, 149(3688):1092–1093, 1965.

[87] JE Welch, FH Harlow, JP Shannon, and BJ Daly. The mac method, los alamos scientific lab. *Report LA-3425, Los Alamos, New Mexico*, 1966.

[88] Francis H Harlow and J Eddie Welch. Numerical study of large-amplitude free-surface motions. *The Physics of Fluids*, 9(5):842–851, 1966.

[89] Francis H Harlow and John P Shannon. The splash of a liquid drop. *Journal of Applied Physics*, 38(10):3855–3866, 1967.

[90] CW Hirt, JL Cook, and TD Butler. A lagrangian method for calculating the dynamics of an incompressible fluid with free surface. *Journal of Computational Physics*, 5(1):103–124, 1970.

[91] Frank A Morrison. Transient multiphase multicomponent flow in porous media. *International Journal of Heat and Mass Transfer*, 16(12):2331–2342, 1973.

[92] James A Viecelli. A computing method for incompressible flows bounded by moving walls. *Journal of Computational Physics*, 8(1):119 – 143, 1971.

[93] CW Hirt, Anthony A Amsden, and JL Cook. An arbitrary lagrangian-eulerian computing method for all flow speeds. *Journal of computational physics*, 14(3):227–253, 1974.

[94] W. F. Noh and Paul Woodward. *SLIC (Simple Line Interface Calculation)*, pages 330–340. Springer Berlin Heidelberg, Berlin, Heidelberg, 1976.

[95] CT Crowe and DE Stock. A computer solution for two-dimensional fluid-particle flows. *International Journal for Numerical Methods in Engineering*, 10(1):185–196, 1976.

[96] Cyril W Hirt and Billy D Nichols. Volume of fluid (vof) method for the dynamics of free boundaries. *Journal of computational physics*, 39(1):201–225, 1981.

[97] D Brian Spalding. Numerical computation of multi-phase fluid flow and heat transfer. In *In Von Karman Inst. for Fluid Dyn. Numerical Computation of Multi-Phase Flows*, 1981.

[98] F. Durst, D. Miloievic, and B. Schönung. Eulerian and lagrangian predictions of particulate two-phase flows: a numerical study. *Applied Mathematical Modelling*, 8(2):101 – 115, 1984.

[99] S Elghobashi. On predicting particle-laden turbulent flows. *Applied scientific research*, 52(4):309–329, 1994.

[100] Paul A Durbin. Separated flow computations with the k-epsilon-v-squared model. *AIAA journal*, 33(4):659–664, 1995.

[101] Todd Arbogast, Steve Bryant, Clint Dawson, Fredrik Saaf, Chong Wang, and Mary Wheeler. Computational methods for multiphase flow and reactive transport problems arising in subsurface contaminant remediation. *Journal of Computational and Applied Mathematics*, 74(1-2):19–32, 1996.

[102] BH Xu and AB Yu. Numerical simulation of the gas-solid flow in a fluidized bed by combining discrete particle method with computational fluid dynamics. *Chemical Engineering Science*, 52(16):2785–2809, 1997.

[103] William J Rider and Douglas B Kothe. Reconstructing volume tracking. *Journal of computational physics*, 141(2):112–152, 1998.

[104] Mark Sussman, Peter Smereka, and Stanley Osher. A level set approach for computing solutions to incompressible two-phase flow. *Journal of Computational physics*, 114(1):146–159, 1994.

[105] Mark Sussman, Emad Fatemi, S Osher, and P Smereka. A level set approach for computing solutions to incompressible two-phase flow ii. Technical report, Lawrence Livermore National Lab., CA (United States), 1995.

[106] JU Brackbill, Douglas B Kothe, and Charles Zemach. A continuum method for modeling surface tension. *Journal of computational physics*, 100(2):335–354, 1992.

[107] Mark Sussman and Elbridge Gerry Puckett. A coupled level set and volume-of-fluid method for computing 3d and axisymmetric incompressible two-phase flows. *Journal of computational physics*, 162(2):301–337, 2000.

[108] DJ Torres and JU Brackbill. The point-set method: front-tracking without connectivity. *Journal of Computational Physics*, 165(2):620–644, 2000.

[109] Douglas Enright, Ronald Fedkiw, Joel Ferziger, and Ian Mitchell. A hybrid particle level set method for improved interface capturing. *Journal of Computational physics*, 183(1):83–116, 2002.

[110] H. Rusche and University of London. *Computational Fluid Dynamics of Dispersed Two-phase Flows at High Phase Fractions*. University of London, 2003.

[111] HG Weller. Derivation, modelling and solution of the conditionally averaged two-phase flow equations. *Nabla Ltd, No Technical Report TR/HGW*, 2, 2002.

[112] Seungwon Shin, SI Abdel-Khalik, Virginie Daru, and Damir Juric. Accurate representation of surface tension using the level contour reconstruction method. *Journal of Computational Physics*, 203(2):493–516, 2005.

[113] D Gerlach, G Tomar, G Biswas, and F Durst. Comparison of volume-of-fluid methods for surface tension-dominant two-phase flows. *International Journal of Heat and Mass Transfer*, 49(3):740–754, 2006.

[114] Francis Weston Sears, Mark Waldo Zemansky, and Hugh D Young. *University physics*. Addison-Wesley, 1987.

[115] Yuehua Yuan and T Randall Lee. Contact angle and wetting properties. In *Surface science techniques*, pages 3–34. Springer, 2013.

[116] Thomas Young. Philosophical transactions of the royal society of london. *Philosophical transactions of the Royal Society of London*, page 119, 1805.

[117] Benny Lautrup. Surface tension. *Physics of Continuous Matter, 2nd edn. CRC Press, Boca Raton*, 2011.

[118] Yulii D Shikhmurzaev. *Capillary flows with forming interfaces*. CRC Press, 2007.

[119] Auro Ashish Saha and Sushanta K Mitra. Effect of dynamic contact angle in a volume of fluid (vof) model for a microfluidic capillary flow. *Journal of colloid and interface science*, 339(2):461–480, 2009.

[120] Serafim Kalliadasis and Hsueh-Chia Chang. Apparent dynamic contact angle of an advancing gas–liquid meniscus. *Physics of Fluids*, 6(1):12–23, 1994.

[121] Š Šikalo, H-D Wilhelm, IV Roisman, S Jakirlić, and C Tropea. Dynamic contact angle of spreading droplets: Experiments and simulations. *Physics of Fluids*, 17(6):062103, 2005.

[122] N Fries and M Dreyer. An analytic solution of capillary rise restrained by gravity. *Journal of colloid and interface science*, 320(1):259–263, 2008.

[123] Zhaoyuan Wang, Jianming Yang, Bonguk Koo, and Frederick Stern. A coupled level set and volume-of-fluid method for sharp interface simulation of plunging breaking waves. *International Journal of Multiphase Flow*, 35(3):227–246, 2009.

[124] A Albadawi, DB Donoghue, AJ Robinson, DB Murray, and YMC Delauré. Influence of surface tension implementation in volume of fluid and coupled volume of fluid with level set methods for bubble growth and detachment. *International Journal of Multiphase Flow*, 53:11–28, 2013.

[125] Ilias Malgarinos, Nikolaos Nikolopoulos, Marco Marengo, Carlo Antonini, and Manolis Gavaises. Vof simulations of the contact angle dynamics during the drop spreading: Standard models and a new wetting force model. *Advances in Colloid and Interface Science*, 212:1 – 20, 2014.

[126] OpenCFDltd. Openfoam. Online, 2017.

[127] George Gabriel Stokes. *On the effect of the internal friction of fluids on the motion of pendulums*, volume 9. Pitt Press Cambridge, 1851.

[128] AB Basset. A treatise on hydrodynamics with numerous examples, volume 2. 1888.

[129] Joseph Boussinesq. *Application dès potentiels a l'étude de l'équilibre et du mouvement des solides élastiques*, volume 4. Gauthier-Villars, 1885.

[130] C.W. Oseen. *Neuere Methoden un d Ergebnisse in der Hydrodynamik.* Akad. Verl.-Ges., 1927.

[131] C.M.Tchen. *Mean value and correlation problems connected with the motion of small particles suspended in a turbulent fluid.* PhD thesis, TU Delft, 1947.

[132] SE Corrsin and J Lumley. On the equation of motion for a particle in turbulent fluid. *Applied Scientific Research*, 6(2):114–116, 1956.

[133] Martin R Maxey and James J Riley. Equation of motion for a small rigid sphere in a nonuniform flow. *The Physics of Fluids*, 26(4):883–889, 1983.

[134] C.T. Crowe, J.D. Schwarzkopf, M. Sommerfeld, and Y. Tsuji. *Multiphase Flows with Droplets and Particles.* Taylor & Francis, 1997.

[135] Inchul Kim, Said Elghobashi, and William A Sirignano. On the equation for spherical-particle motion: effect of reynolds and acceleration numbers. *Journal of Fluid Mechanics*, 367:221–253, 1998.

[136] Daniel J Brandon and SK Aggarwal. A numerical investigation of particle deposition on a square cylinder placed in a channel flow. *Aerosol Science & Technology*, 34(4):340–352, 2001.

[137] VDI Gesellschaft. *VDI Heat Atlas.* VDI-Buch. Springer Berlin Heidelberg, 2010.

[138] A Shirvanian, M Faghri, Z Zhang, and Y Asako. Numerical solution of the effect of vibration on melting of unfixed rectangular phase-change material under variable-gravity environment. *Numerical Heat Transfer, Part A Applications*, 34(3):257–278, 1998.

[139] Essam H Atta and Joseph Vadyak. A grid overlapping scheme for flowfield computations about multicomponent configurations. *AIAA Journal(ISSN 0001-1452)*, 21:1271–1277, 1983.

[140] Göran Starius. On composite mesh difference methods for hyperbolic differential equations. *Numerische Mathematik*, 35(3):241–255, 1980.

[141] E. H. Atta and J. Vadyak. *A grid interfacing zonal algorithm for three-dimensional transonic flows about aircraft configurations*, pages 107–114. Springer Berlin Heidelberg, Berlin, Heidelberg, 1982.

[142] Marsha J Berger and Joseph Oliger. Adaptive mesh refinement for hyperbolic partial differential equations. *Journal of computational Physics*, 53(3):484–512, 1984.

[143] William Douglas Henshaw. *Part I. The numerical solution of hyperbolic systems of conservation laws. Part II. Composite overlapping grid techniques.* PhD thesis, California Institute of Technology, 1985.

[144] C Wayne Mastin. Interface procedures for overlapping grids. Technical report 19860008591, National Aeronautics and Space Administration (NASA), USA, January 1985.

[145] Gerald L Browning, James J Hack, and Paul N Swarztrauber. A comparison of three numerical methods for solving differential equations on the sphere. *Monthly Weather Review*, 117(5):1058–1075, 1989.

[146] G Chesshire and William D Henshaw. Composite overlapping meshes for the solution of partial differential equations. *Journal of Computational Physics*, 90(1):1–64, 1990.

[147] Laszlo Fuchs. Calculation of flow fields using overlapping grids. In *Proceedings of the Eighth GAMM-Conference on Numerical Methods in Fluid Mechanics*, pages 138–147. Springer, 1990.

[148] JY Tu and L Fuchs. Overlapping grids and multigrid methods for three-dimensional unsteady flow calculations in ic engines. *International journal for numerical methods in fluids*, 15(6):693–714, 1992.

[149] Peter K Moore and Joseph E Flaherty. Adaptive local overlapping grid methods for parabolic systems in two space dimensions. *Journal of Computational Physics*, 98(1):54–63, 1992.

[150] Robert L Meakin. Moving body overset grid methods for complete aircraft tiltrotor simulations. *AIAA paper*, 93:3350, 1993.

[151] William D Henshaw. A fourth-order accurate method for the incompressible navier-stokes equations on overlapping grids. *Journal of computational physics*, 113(1):13–25, 1994.

[152] Kai-Hsiung Kao and Meng-Sing Liou. Application of chimera/unstructured hybrid grids for conjugate heat transfer. *AIAA journal*, 35(9), 1997.

[153] William D Henshaw. Overture: An object-oriented framework for overlapping grid applications. In *AIAA conference on Applied Aerodynamics*, 2002.

[154] Aziz Azimi, Siamak Kazemzadeh Hannani, and Bijan Farhanieh. Structured multiblock body-fitted grids solution of transient inverse heat conduction problems in an arbitrary geometry. *Numerical Heat Transfer, Part B: Fundamentals*, 54(3):260–290, 2008.

[155] William D Henshaw and Kyle K Chand. A composite grid solver for conjugate heat transfer in fluid–structure systems. *Journal of Computational Physics*, 228(10):3708–3741, 2009.

[156] Luca Mangani, Matteo Cerutti, Massimiliano Maritano, and Martin Spel. Conjugate heat transfer analysis of nasa c3x film cooled vane with an object-oriented cfd code. *ASME Paper No. GT2010-23458*, pages 1805–1814, 2010.

[157] Henrik Rusche and Hrvoje Jasak. Implicit solution techniques for coupled multifield problems block solution, coupled matrices. In *Training at the OpenFOAM Workshop, Gothenburg, Sweden*, 2010.

[158] J Geertsma et al. The effect of fluid pressure decline on volumetric changes of porous rocks. 1957.

[159] Rodney Hill. Elastic properties of reinforced solids: some theoretical principles. *Journal of the Mechanics and Physics of Solids*, 11(5):357–372, 1963.

[160] R.M. Christensen. Viscoelastic properties of heterogeneous media. *Journal of the Mechanics and Physics of Solids*, 17(1):23 – 41, 1969.

[161] BW Rosen. Thermomechanical properties of fibrous composites. In *Proceedings of the Royal Society of London A: Mathematical, Physical and Engineering Sciences*, volume 319, pages 79–94. The Royal Society, 1970.

[162] Rodney Hill. On constitutive macro-variables for heterogeneous solids at finite strain. In *Proceedings of the Royal Society of London A: mathematical, physical and engineering sciences*, volume 326, pages 131–147. The Royal Society, 1972.

[163] Shlomo P Neuman. Theoretical derivation of darcy's law. *Acta Mechanica*, 25(3-4):153–170, 1977.

[164] Warren A. Hall. An analytical derivation of the darcy equation. *Eos, Transactions American Geophysical Union*, 37(2):185–188, 1956.

[165] M. KING HUBBERT. Darcy's law and the field equations of the flow of underground fluids. *International Association of Scientific Hydrology. Bulletin*, 2(1):23–59, 1957.

[166] J. Bear. *Dynamics of Fluids in Porous Media*. Dover Civil and Mechanical Engineering Series. Dover, 1972.

[167] P De Buhan and A Taliercio. A homogenization approach to the yield strength of composite materials. *European journal of mechanics. A. Solids*, 10(2):129–154, 1991.

[168] Majid Hassanizadeh and William G Gray. General conservation equations for multi-phase systems: 1. averaging procedure. *Advances in water resources*, 2:131–144, 1979.

[169] S Majid Hassanizadeh and William G Gray. High velocity flow in porous media. *Transport in porous media*, 2(6):521–531, 1987.

[170] S Liu and JH Masliyah. Continuum approach to single phase flow in porous media. *WIT Transactions on Engineering Sciences*, 9, 1970.

[171] Stephen Whitaker. Simultaneous heat, mass, and momentum transfer in porous media: a theory of drying. *Advances in heat transfer*, 13:119–203, 1977.

[172] Stephen Whitaker. Flow in porous media ii: The governing equations for immiscible, two-phase flow. *Transport in porous media*, 1(2):105–125, 1986.

[173] Stephen Whitaker. Local thermal equilibrium: an application to packed bed catalytic reactor design. *Chemical Engineering Science*, 41(8):2029–2039, 1986.

[174] OA Plumb and S Whitaker. Dispersion in heterogeneous porous media: 1. local volume averaging and large-scale averaging. *Water Resources Research*, 24(7):913–926, 1988.

[175] A Amiri and K Vafai. Analysis of dispersion effects and non-thermal equilibrium, non-darcian, variable porosity incompressible flow through porous media. *International Journal of Heat and Mass Transfer*, 37(6):939–954, 1994.

[176] Sabri Ergun. Fluid flow through packed columns. *Chem. Eng. Prog.*, 48:89–94, 1952.

[177] N Wakao and T Funazkri. Effect of fluid dispersion coefficients on particle-to-fluid mass transfer coefficients in packed beds: correlation of sherwood numbers. *Chemical Engineering Science*, 33(10):1375–1384, 1978.

[178] N Wakao, S Kaguei, and T Funazkri. Effect of fluid dispersion coefficients on particle-to-fluid heat transfer coefficients in packed beds: correlation of nusselt numbers. *Chemical engineering science*, 34(3):325–336, 1979.

[179] A. Amiri and K. Vafai. Transient analysis of incompressible flow through a packed bed. *International Journal of Heat and Mass Transfer*, 41(24):4259 – 4279, 1998.

[180] KAR Ismail and R Stuginsky Jr. A parametric study on possible fixed bed models for pcm and sensible heat storage. *Applied Thermal Engineering*, 19(7):757–788, 1999.

[181] KAR Ismail and MM Gongalves. Effectiveness ntu performance of finned pcm storage unit. *WIT Transactions on Modelling and Simulation*, 6, 1970.

[182] V Ananthanarayanan, Y Sahai, CE Mobley, and RA Rapp. Modeling of fixed bed heat storage units utilizing phase change materials. *Metallurgical Transactions B*, 18(2):339–346, 1987.

[183] Donald E Beasley and John A Clark. Transient response of a packed bed for thermal energy storage. *International Journal of Heat and Mass Transfer*, 27(9):1659–1669, 1984.

[184] Osama Mesalhy, Khalid Lafdi, Ahmed Elgafy, and Keith Bowman. Numerical study for enhancing the thermal conductivity of phase change material (pcm) storage using high thermal conductivity porous matrix. *Energy Conversion and Management*, 46(6):847–867, 2005.

[185] M. Sözen and K. Vafai. Analysis of the non-thermal equilibrium condensing flow of a gas through a packed bed. *International Journal of Heat and Mass Transfer*, 33(6):1247–1261, 1990.

[186] Bader Alazmi and Kambiz Vafai. Constant wall heat flux boundary conditions in porous media under local thermal non-equilibrium conditions. *International Journal of Heat and Mass Transfer*, 45(15):3071–3087, 2002.

[187] K.A.R. Ismail and J.R. Henrıquez. Numerical and experimental study of spherical capsules packed bed latent heat storage system. *Applied Thermal Engineering*, 22(15):1705 – 1716, 2002.

[188] C. Arkar and S. Medved. Influence of accuracy of thermal property data of a phase change material on the result of a numerical model of a packed bed latent heat storage with spheres. *Thermochimica Acta*, 438(1):192 – 201, 2005.

[189] A. Felix Regin, S.C. Solanki, and J.S. Saini. An analysis of a packed bed latent heat thermal energy storage system using pcm capsules: Numerical investigation. *Renewable Energy*, 34(7):1765 – 1773, 2009.

[190] L Xia, P Zhang, and RZ Wang. Numerical heat transfer analysis of the packed bed latent heat storage system based on an effective packed bed model. *Energy*, 35(5):2022–2032, 2010.

[191] Giw Zanganeh, Andrea Pedretti, Simone Zavattoni, Maurizio Barbato, and Aldo Steinfeld. Packed-bed thermal storage for concentrated solar power–pilot-scale demonstration and industrial-scale design. *Solar Energy*, 86(10):3084–3098, 2012.

[192] Hitesh Bindra, Pablo Bueno, Jeffrey F Morris, and Reuel Shinnar. Thermal analysis and exergy evaluation of packed bed thermal storage systems. *Applied Thermal Engineering*, 52(2):255–263, 2013.

[193] Christopher J. Greenshields. *OpenFOAM Programmers guide*. CFD Direct Ltd., OpenFOAM Foundation Ltd., 2.2 edition, 2013. licensed under a Creative Commons Attribution-NonCommercial-NoDerivs 2.2 Unported License.

[194] S.B. Pope. *Turbulent Flows*. Cambridge University Press, August 2000. ISBN: 0521598869, 9780521598866.

[195] Odd Faltinsen. *Sea loads on ships and offshore structures*, volume 1. Cambridge university press, 1993.

[196] Cengel. *Heat & Mass Transfer: A Practical Approach*. McGraw-Hill Education (India) Pvt Limited, 2007.

[197] Mohammed Farid, YONGSIK KIM, TAKUYA HONDA, and ATSUSHI KANZAWA. The role of natural convection during melting and solidification of pcm in a vertical cylinder. *Chemical Engineering Communications*, 84(1):43–60, 1989.

[198] MM Farid and AK Mohamed. Effect of natural convection on the process of melting and solidification of paraffin wax. *Chemical engineering communications*, 57(1-6):297–316, 1987.

[199] R.H.Marshall. Natural convection effect in rectangular enclosers containing a phase change material. In *Thermal Storage and Heat Transfer in Solar Energy System*, pages 61–69. ASME, New York, 1970.

[200] Mohammed M. Farid and Rafah M. Husian. An electrical storage heater using the phase-change method of heat storage. *Energy Conversion and Management*, 30(3):219 – 230, 1990.

[201] P Wang, K Vafai, DY Liu, and C Xu. Analysis of collimated irradiation under local thermal non-equilibrium condition in a packed bed. *International Journal of Heat and Mass Transfer*, 80:789–801, 2015.

[202] Heinz Brauer. *Grundlagen der Einphasen-und Mehrphasenströmungen*, volume 2. Sauerländer, 1971.

[203] Warren M Rohsenow, James P Hartnett, Young I Cho, et al. *Handbook of heat transfer*, volume 3. McGraw-Hill New York, 1998.

[204] ICONFINDER. https://www.iconfinder.com/icons/1287170/camera-dslr-photo-photography-picture-side-view-icon, 2017. Paid.

[205] LAUDA DR. R. WOBSERGMBH & CO. KG. Lauda operating instructions. Online, 97912 Lauda-Koenigshofen. Srie Z 01 01/01 YACE 0050.

[206] BEHA. *Bedienungsanleitung Labornetzgeräte UNIWATT NG-310*. PEWA Messtechnik GmbH, Schwerte. (in German).

[207] ICONFINDER. https://www.iconfinder.com/icons/2351543/camera-dslr-lens-photo-photography-icon-size-128, 2017. Paid.

[208] Rubitherm. *Technisches Datenblatt RT35HC*. Rubitherm Technologies GmbH, D-12277, Berlin, Januar 2015. (in German).

[209] Huber. Huber betriebsanleitung. Online, Offenburg, August 2017. (in German).

[210] Iconfinder. Computer laptop icon 103832. Online, August 2017.

[211] Stephan Höhlein, Andreas König-Haagen, and Dieter Brüggemann. Thermophysical characterization of mgcl2· 6h2o, xylitol and erythritol as phase change materials (pcm) for latent heat thermal energy storage (lhtes). *Materials*, 10(4):444, 2017.

[212] D. Brüggemann. *Entwicklung makroverkapselter Latentwärmespeicher für den straßengebundenen Transport von Abwärme (MALATrans)*. Universität Bayreuth, 2017.

[213] P.A. Galione, O. Lehmkuhl, J. Rigola, and A. Oliva. Fixed-grid numerical modeling of melting and solidification using variable thermo-physical properties – application to the melting of n-octadecane inside a spherical capsule. *International Journal of Heat and Mass Transfer*, 86(Supplement C):721 – 743, 2015.

[214] Sebastian Pinnau. Strömungs- und kältetechnische optimierung von latentkältespeichern. Abschlussbericht Az: 24825, Technical University Dresden, Dresden, 2009.

Vorveröffentlichungen

- Kasibhatla, Raghavendra Rohith, Andreas König-Haagen, Fabian Rösler and Dieter Brüggemann. "Numerical modelling of melting and settling of an encapsulated PCM using variable viscosity." Heat and Mass Transfer 53.5 (2017): 1735-1744.
- Kasibhatla, Raghavendra Rohith, Andreas König-Haagen and Dieter Brüggemann. "Numerical modelling of wetting phenomena during melting of PCM." Procedia Engineering 157 (2016): 139-147.
- Kasibhatla, Raghavendra Rohith, Andreas Konig-Haagen, and Dieter Brüggemann. "Numerical study of convective term for a solid-liquid phase change problem." Multiphase Science and Technology 30, no. 2-3 (2018).

In der Reihe „*Thermodynamik: Energie, Umwelt, Technik*“, herausgegeben von Prof. Dr.-Ing. D. Brüggemann, bisher erschienen:

ISSN 1611-8421

1	Dietmar Zeh	Entwicklung und Einsatz einer kombinierten Raman/Mie-Streulichtmesstechnik zur ein- und zweidimensionalen Untersuchung der dieselmotorischen Gemischbildung	
		ISBN 978-3-8325-0211-9	40.50 €
2	Lothar Herrmann	Untersuchung von Tropfengrößen bei Injektoren für Ottomotoren mit Direkteinspritzung	
		ISBN 978-3-8325-0345-1	40.50 €
3	Klaus-Peter Gansert	Laserinduzierte Tracerfluoreszenz-Untersuchungen zur Gemischaufbereitung am Beispiel des Ottomotors mit Saugrohreinspritzung	
		ISBN 978-3-8325-0362-8	40.50 €
4	Wolfram Kaiser	Entwicklung und Charakterisierung metallischer Bipolarplatten für PEM-Brennstoffzellen	
		ISBN 978-3-8325-0371-0	40.50 €
5	Joachim Boltz	Orts- und zyklusaufgelöste Bestimmung der Rußkonzentration am seriennahen DI-Dieselmotor mit Hilfe der Laserinduzierten Inkandeszenz	
		ISBN 978-3-8325-0485-4	40.50 €
6	Hartmut Sauter	Analysen und Lösungsansätze für die Entwicklung von innovativen Kurbelgehäuseentlüftungen	
		ISBN 978-3-8325-0529-5	40.50 €
7	Cosmas Heller	Modellbildung, Simulation und Messung thermofluiddynamischer Vorgänge zur Optimierung des Flowfields von PEM-Brennstoffzellen	
		ISBN 978-3-8325-0675-9	40.50 €

8	Bernd Mewes	Entwicklung der Phasenspezifischen Raman-Spektroskopie zur Untersuchung der Gemischbildung in Methanol- und Ethanolsprays ISBN 978-3-8325-0841-8 40.50 €
9	Tobias Schittkowski	Laserinduzierte Inkandeszenz an Nanopartikeln ISBN 978-3-8325-0887-6 40.50 €
10	Marc Schröter	Einsatz der SMPS- und Lidar-Messtechnik zur Charakterisierung von Nanopartikeln und gasförmigen Verunreinigungen in der Luft ISBN 978-3-8325-1275-0 40.50 €
11	Stefan Hildenbrand	Simulation von Verdampfungs- und Vermischungsvorgängen bei motorischen Sprays mit Wandaufprall ISBN 978-3-8325-1428-0 40.50 €
12	Stefan Staudacher	Numerische Simulation turbulenter Verbrennungsvorgänge in Raketenbrennkammern bei Einbringen flüssigen Sauerstoffs ISBN 978-3-8325-1430-3 40.50 €
13	Dang Cuong Phan	Optimierung thermischer Prozesse bei der Glasproduktion durch Modellierung und Simulation ISBN 978-3-8325-1773-1 40.50 €
14	Ulli Drescher	Optimierungspotenzial des Organic Rankine Cycle für biomassebefeuerte und geothermische Wärmequellen ISBN 978-3-8325-1912-4 39.00 €
15	Martin Gallinger	Kaltstart- und Lastwechselverhalten der Onboard-Wasserstofferzeugung durch katalytische partielle Oxidation für Brennstoffzellenfahrzeuge ISBN 978-3-8325-1915-5 40.50 €

16 Kai Gartung

Modellierung der Verdunstung realer Kraftstoffe zur Simulation der Gemischbildung bei Benzindirekteinspritzung

ISBN 978-3-8325-1934-6 40.50 €

17 Luis Matamoros

Numerical Modeling and Simulation of PEM Fuel Cells under Different Humidifying Conditions

ISBN 978-3-8325-2174-5 34.50 €

18 Eva Schießwohl

Entwicklung eines Kaltstartkonzeptes für ein Polymermembran-Brennstoffzellensystem im automobilen Einsatz

ISBN 978-3-8325-2450-0 36.50 €

19 Christian Hüttl

Einfluss der Sprayausbreitung und Gemischbildung auf die Verbrennung von Biodiesel-Diesel-Gemischen

ISBN 978-3-8325-3009-9 37.00 €

20 Salih Manasra

Combustion and In-Cylinder Soot Formation Characteristics of a Neat GTL-Fueled DI Diesel Engine

ISBN 978-3-8325-3001-3 36.50 €

21 Dietmar Böker

Laserinduzierte Plasmen zur Zündung von Wasserstoff-Luft-Gemischen bei motorrelevanten Drücken

ISBN 978-3-8325-3036-5 43.50 €

22 Florian Heberle

Untersuchungen zum Einsatz von zeotropen Fluidgemischen im Organic Rankine Cycle für die geothermische Stromerzeugung

ISBN 978-3-8325-3355-7 39.00 €

23 Ulrich Leidenberger

Optische und analytische Untersuchungen zum Einfluss dieselmotorischer Parameter auf die physikochemischen Eigenschaften emittierter Rußpartikel

ISBN 978-3-8325-3616-9 43.00 €

24	Fabian Rösler	Modellierung und Simulation der Phasenwechselvorgänge in makroverkapselten latenten thermischen Speichern ISBN 978-3-8325-3787-6 44.50 €
25	Markus Preißinger	Thermoökonomische Bewertung des Organic Rankine Cycles bei der Stromerzeugung aus industrieller Abwärme ISBN 978-3-8325-3866-8 41.00 €
26	Sebastian Lorenz	Impulskettenzündung mit passiv gütegeschalteten Laserzündkerzen unter motorischen Bedingungen ISBN 978-3-8325-4235-1 65.00 €
27	Sebastian Lehmann	Optische Untersuchung der Verdunstung von Kraftstofftropfen bei niedrigen Umgebungstemperaturen ISBN 978-3-8325-4433-1 53.00 €
28	Wolfgang Mühlbauer	Analyse der Gemischbildung, der Verbrennung und der Partikelemissionen von alternativen Dieselkraftstoffen mit optischen und analytischen Messmethoden ISBN 978-3-8325-4516-1 59.00 €
29	Theresa Weith	Wärmeübergang beim Sieden von linearen Siloxanen und Siloxangemischen im horizontalen Rohr ISBN 978-3-8325-4601-4 43.00 €
30	Andreas König-Haagen	Exergetische Bewertung thermischer Speicher in Systemen am Beispiel einer Anlage mit kombinierter Stromerzeugung ISBN 978-3-8325-4876-6 51.50 €
31	Christian Pötzinger	Modellierung und thermoökonomische Bewertung von PV-Energiesystemen mit Batterie- und Wasserstoffspeichern in Wohnhäusern ISBN 978-3-8325-4878-0 58.00 €

32	Mark Bärwinkel	Einfluss der Fokussierung und der Impulsenergie auf die Entflammung bei der Zündung mit einem passiv gütegeschalteten Laser ISBN 978-3-8325-4882-7 52.50 €
33	Raghavendra Rohith Kasibhatla	Multiscale thermo-fluid modelling of macro-encapsulated latent heat thermal energy storage systems ISBN 978-3-8325-4893-3 52.00 €